NICHOLAS CARR

Nicholas Carr is the author of *The Shallows: What the Internet Is Doing to Our Brains*, a 2011 Pulitzer Prize nominee and a *New York Times* bestseller, as well as two other influential books, *The Big Switch: Rewiring the World, from Edison to Google* (2008) and *Does IT Matter?* (2004). His books have been translated into more than 20 languages. www.nicholascarr.com

NICHOLAS CARR

The Glass Cage

Who Needs Humans Anyway?

VINTAGE

1 3 5 7 9 10 8 6 4 2

Vintage
20 Vauxhall Bridge Road,
London SW1V 2SA

Vintage is part of the Penguin Random House group of companies
whose addresses can be found at global.penguinrandomhouse.com.

 Penguin
Random House
UK

First published in Vintage in 2016
First published in hardback by The Bodley Head in 2015

www.vintage-books.co.uk

A CIP catalogue record for this book
is available from the British Library

ISBN 9780099597452

Printed and bound by CPI Group (UK) Ltd, Croydon CR0 4YY

 MIX
Paper from
responsible sources
FSC® C018179

Penguin Random House is committed to a sustainable future
for our business, our readers and our planet. This book is
made from Forest Stewardship Council® certified paper

To Ann

CONTENTS

No one

to witness

and adjust, no one to drive the car

—*William Carlos Williams*

THE GLASS CAGE

ALERT FOR OPERATORS

ON JANUARY 4, 2013, THE FIRST FRIDAY OF A NEW YEAR, a dead day newswise, the Federal Aviation Administration released a one-page notice. It had no title. It was identified only as a "safety alert for operators," or SAFO. Its wording was terse and cryptic. In addition to being posted on the FAA's website, it was sent to all U.S. airlines and other commercial air carriers. "This SAFO," the document read, "encourages operators to promote manual flight operations when appropriate." The FAA had collected evidence, from crash investigations, incident reports, and cockpit studies, indicating that pilots had become too dependent on autopilots and other computerized systems. Overuse of flight automation, the agency warned, could "lead to degradation of the pilot's ability to quickly recover the aircraft from an undesired state." It could, in blunter terms, put a plane and its passengers in jeopardy. The alert concluded with a recommendation that airlines, as a matter of operational policy, instruct pilots to spend less time flying on autopilot and more time flying by hand.[1]

This is a book about automation, about the use of computers and software to do things we used to do ourselves. It's not about the technology or the economics of automation, nor is it about the future of robots and cyborgs and gadgetry, though all those things enter into the story. It's about automation's human consequences. Pilots have been out in front of a wave that is now engulfing us. We're looking to computers to shoulder more of our work, on the job and off, and to guide us through more of our everyday routines. When we need to get something done today, more often than not we sit down in front of a monitor, or open a laptop, or pull out a smartphone, or strap a net-connected accessory to our forehead or wrist. We run apps. We consult screens. We take advice from digitally simulated voices. We defer to the wisdom of algorithms.

Computer automation makes our lives easier, our chores less burdensome. We're often able to accomplish more in less time—or to do things we simply couldn't do before. But automation also has deeper, hidden effects. As aviators have learned, not all of them are beneficial. Automation can take a toll on our work, our talents, and our lives. It can narrow our perspectives and limit our choices. It can open us to surveillance and manipulation. As computers become our constant companions, our familiar, obliging helpmates, it seems wise to take a closer look at exactly how they're changing what we do and who we are.

CHAPTER ONE

PASSENGERS

AMONG THE HUMILIATIONS OF MY TEENAGE YEARS WAS ONE that might be termed psycho-mechanical: my very public struggle to master a manual transmission. I got my driver's license early in 1975, not long after I turned sixteen. The previous fall, I had taken a driver's ed course with a group of my high-school classmates. The instructor's Oldsmobile, which we used for our on-the-road lessons and then for our driving tests at the dread Department of Motor Vehicles, was an automatic. You pressed the gas pedal, you turned the wheel, you hit the brakes. There were a few tricky maneuvers—making a three-point turn, backing up in a straight line, parallel parking—but with a little practice among pylons in the school parking lot, even they became routine.

License in hand, I was ready to roll. There was just one last road-block. The only car available to me at home was a Subaru sedan with a stick shift. My dad, not the most hands-on of parents, granted me a single lesson. He led me out to the garage one Saturday morning, plopped himself down behind the wheel, and had me climb into the passenger seat beside him. He placed my left palm over the shift knob and guided my hand through the gears: "That's first." Brief pause. "Second." Brief pause. "Third." Brief pause. "Fourth." Brief pause. "Down over here"—a pain shot through my wrist as it twisted into an unnatural position—"is Reverse." He glanced at me to confirm I had it all down. I nodded helplessly. "And that"—wiggling my hand back and forth—"that's Neutral." He gave me a few tips about the speed ranges of the four forward gears. Then he pointed to the clutch pedal he had pinned beneath his loafer. "Make sure you push that in while you shift."

I proceeded to make a spectacle of myself on the roads of the small New England town where we lived. The car would buck as I tried to find the correct gear, then lurch forward as I mistimed the release of the clutch. I'd stall at every red light, then stall again halfway out into the intersection. Hills were a horror. I'd let the clutch out too quickly, or too slowly, and the car would roll backward until it came to rest against the bumper of the vehicle behind me. Horns were honked, curses cursed, birds flipped. What made the experience all the more excruciating was the Subaru's yellow paint job—the kind of yellow you get with a kid's rain slicker or a randy male goldfinch. The car was an eye magnet, my flailing impossible to miss.

From my putative friends, I received no sympathy. They found my struggles a source of endless, uproarious amusement. "Grind me a pound!" one of them would yell with glee from the backseat whenever I'd muff a shift and set off a metallic gnashing of gear teeth. "Smooth move," another would snigger as the engine rattled to a stall. The word "spaz"—this was well before anyone had heard of political correctness—was frequently lobbed my way. I had a suspicion that my incompetence with the stick was something my buddies laughed about behind my back. The metaphorical implications were not lost on me. My manhood, such as it was at sixteen, felt deflated.

But I persisted—what choice did I have?—and after a week or two I began to get the hang of it. The gearbox loosened up and became more forgiving. My arms and legs stopped working at cross-purposes and started cooperating. Soon, I was shifting without thinking about it. It just happened. The car no longer stalled or bucked or lurched. I no longer had to sweat the hills or the intersections. The transmission and I had become a team. We meshed. I took a quiet pride in my accomplishment.

Still, I coveted an automatic. Although stick shifts were fairly common back then, at least in the econoboxes and junkers that kids drove, they had already taken on a behind-the-times, hand-me-down quality. They seemed fusty, a little yesterday. Who wanted to be "manual" when you could be "automatic"? It was like the difference between scrubbing dishes by hand and sticking them in a dishwasher. As it turned out, I didn't have to wait long for my wish to be granted. Two years after I got my license, I managed to total the Subaru during a late-night misadventure, and not long afterward I took stewardship of a used, cream-colored, two-door Ford Pinto. The car was a piece of crap—some now see the Pinto as marking the nadir of American manufacturing in the twentieth century—but to me it was redeemed by its automatic transmission.

I was a new man. My left foot, freed from the demands of the clutch, became an appendage of leisure. As I tooled around town, it would sometimes tap along jauntily to the thwacks of Charlie Watts or the thuds of John Bonham—the Pinto also had a built-in eight-track deck, another touch of modernity—but more often than not it just stretched out in its little nook under the left side of the dash and napped. My right hand became a beverage holder. I not only felt renewed and up-to-date. I felt liberated.

It didn't last. The pleasures of having less to do were real, but they faded. A new emotion set in: boredom. I didn't admit it to anyone, hardly to myself even, but I began to miss the gear stick and the clutch pedal. I missed the sense of control and involvement they had given me—the ability to rev the engine as high as I wanted, the feel of the clutch releasing and the gears grabbing, the tiny thrill that came with a downshift at speed. The automatic made me feel a little less like a driver and a little more like a passenger. I came to resent it.

■   ■   ■   ■

MOTOR AHEAD thirty-five years, to the morning of October 9, 2010. One of Google's in-house inventors, the German-born roboticist Sebastian Thrun, makes an extraordinary announcement in a blog post. The company has developed "cars that can drive themselves." These aren't some gawky, gearhead prototypes puttering around the Googleplex's parking lot. These are honest-to-goodness street-legal vehicles—Priuses, to be precise—and, Thrun reveals, they've already logged more than a hundred thousand miles on roads and highways in California and Nevada. They've cruised down Holly-wood Boulevard and the Pacific Coast Highway, gone back and forth over the Golden Gate Bridge, circled Lake Tahoe. They've merged into freeway traffic, crossed busy intersections, and inched through rush-hour gridlock. They've swerved to avoid collisions. They've done all this by themselves. Without human help. "We think this is a first in robotics research," Thrun writes, with sly humility.[1]

Building a car that can drive itself is no big deal. Engineers and tinkerers have been constructing robotic and remote-controlled auto-mobiles since at least the 1980s. But most of them were crude jalop-ies. Their use was restricted to test-drives on closed tracks or to races and rallies in deserts and other remote areas, far away from pedes-trians and police. The Googlemobile, Thrun's announcement made clear, is different. What makes it such a breakthrough, in the history of both transport and automation, is its ability to navigate the real world in all its chaotic, turbulent complexity. Outfitted with laser range-finders, radar and sonar transmitters, motion detectors, video cameras, and GPS receivers, the car can sense its surroundings in minute detail. It can see where it's going. And by processing all the streams of incoming information instantaneously—in "real time"— its onboard computers are able to work the accelerator, the steering wheel, and the brakes with the speed and sensitivity required to drive on actual roads and respond fluidly to the unexpected events that drivers always encounter. Google's fleet of self-driving cars has now racked up close to a million miles, and the vehicles have caused just one serious accident. That was a five-car pileup near the company's Silicon Valley headquarters in 2011, and it doesn't really count. It happened, as Google was quick to announce, "while a person was manually driving the car."[2]

Autonomous automobiles have a ways to go before they start chauffeuring us to work or ferrying our kids to soccer games. Although Google has said it expects commercial versions of its car to be on sale by the end of the decade, that's probably wishful thinking. The vehicle's sensor systems remain prohibitively expensive, with the roof-mounted laser apparatus alone going for eighty thousand dollars. Many technical challenges remain to be met, such as navigating snowy or leaf-covered roads, dealing with unexpected detours, and interpreting the hand signals of traffic cops and road workers. Even the most powerful computers still have a hard time distinguishing a bit of harmless road debris (a flattened cardboard box, say) from a dangerous obstacle (a nail-studded chunk of plywood). Most daunting of all are the many legal, cultural, and ethical hurdles a driverless car faces. Where, for instance, will culpability and liability reside should a computer-driven automobile cause an accident that kills or injures someone? With the car's owner? With the manufacturer that installed the self-driving system? With the programmers who wrote the software? Until such thorny questions get sorted out, fully automated cars are unlikely to grace dealer showrooms.

Progress will sprint forward nonetheless. Much of the Google test cars' hardware and software will come to be incorporated into future generations of cars and trucks. Since the company went public with its autonomous vehicle program, most of the world's major carmakers have let it be known that they have similar efforts under way. The goal, for the time being, is not so much to create an immaculate robot-on-wheels as to continue to invent and refine automated features that enhance safety and convenience in ways that get people to buy new cars. Since I first turned the key in my Subaru's ignition, the automation of driving has already come a long way. Today's automobiles are stuffed with electronic gadgetry. Microchips and sensors govern the workings of the cruise control, the antilock brakes, the traction and stability mechanisms, and, in higher-end models, the variable-speed transmission, parking-assist system, collision-avoidance system, adaptive headlights, and dashboard displays. Software already provides a buffer between us and the road. We're not so much controlling our cars as sending electronic inputs to the computers that control them.

In coming years, we'll see responsibility for many more aspects of driving shift from people to software. Luxury-car makers like Infiniti, Mercedes, and Volvo are rolling out models that combine radar-assisted adaptive cruise control, which works even in stop-and-go traffic, with computerized steering systems that keep a car centered in its lane and brakes that slam themselves on in emergencies. Other manufacturers are rushing to introduce even more advanced controls. Tesla Motors, the electric car pioneer, is developing an automotive autopilot that "should be able to [handle] 90 percent of miles driven," according to the company's ambitious chief executive, Elon Musk.[3]

The arrival of Google's self-driving car shakes up more than our conception of driving. It forces us to change our thinking about what computers and robots can and can't do. Up until that fateful October day, it was taken for granted that many important skills lay beyond the reach of automation. Computers could do a lot of things, but they couldn't do everything. In an influential 2004 book, *The New Division of Labor: How Computers Are Creating the Next Job Market*, economists Frank Levy and Richard Murnane argued, convincingly, that there were practical limits to the ability of software programmers to replicate human talents, particularly those involving sensory perception, pattern recognition, and conceptual knowledge. They pointed specifically to the example of driving a car on the open road, a talent that requires the instantaneous interpretation of a welter of visual signals and an ability to adapt seamlessly to shifting and often unanticipated situations. We hardly know how we pull off such a feat ourselves, so the idea that programmers could reduce all of driving's intricacies, intangibilities, and contingencies to a set of instructions, to lines of software code, seemed ludicrous. "Executing a left turn across oncoming traffic," Levy and Murnane wrote, "involves so many factors that it is hard to imagine the set of rules that can replicate a driver's behavior." It seemed a sure bet, to them and to pretty much everyone else, that steering wheels would remain firmly in the grip of human hands.[4]

In assessing computers' capabilities, economists and psychologists have long drawn on a basic distinction between two kinds of knowledge: *tacit* and *explicit*. Tacit knowledge, which is also sometimes called procedural knowledge, refers to all the stuff we do without thinking about it: riding a bike, snagging a fly ball, reading a book, driving a car. These aren't innate skills—we have to learn them, and some people are better at them than others—but they can't be expressed as a simple recipe. When you make a turn through a busy intersection in your car, neurological studies show, many areas of your brain are hard at work, processing sensory stimuli, making estimates of time and distance, and coordinating your arms and legs.[5] But if someone asked you to document everything involved in making that turn, you wouldn't be able to, at least not without resorting to generalizations and abstractions. The ability resides deep in your nervous system, outside the ambit of your conscious mind. The mental processing goes on without your awareness.

Much of our ability to size up situations and make quick judgments about them stems from the fuzzy realm of tacit knowledge. Most of our creative and artistic skills reside there too. Explicit knowledge, which is also known as declarative knowledge, is the stuff you can actually write down: how to change a flat tire, how to fold an origami crane, how to solve a quadratic equation. These are processes that can be broken down into well-defined steps. One person can explain them to another person through written or oral instructions: do this, then this, then this.

Because a software program is essentially a set of precise, written instructions—do this, then this, then this—we've assumed that while computers can replicate skills that depend on explicit knowledge, they're not so good when it comes to skills that flow from tacit knowledge. How do you translate the ineffable into lines of code, into the rigid, step-by-step instructions of an algorithm? The boundary between the explicit and the tacit has always been a rough one—a lot of our talents straddle the line—but it seemed to offer a good way to define the limits of automation and, in turn, to mark out the exclusive precincts of the human. The sophisticated jobs Levy and Murnane identified as lying beyond the reach of computers—in addition to driving, they pointed to teaching and medical diagnosis—were a mix of the mental and the manual, but they all drew on tacit knowledge.

Google's car resets the boundary between human and computer, and it does so more dramatically, more decisively, than have earlier breakthroughs in programming. It tells us that our idea of the limits of automation has always been something of a fiction. We're not as special as we think we are. While the distinction between tacit and explicit knowledge remains a useful one in the realm of human psychology, it has lost much of its relevance to discussions of automation.

*   *   *   *

THAT DOESN'T mean that computers now have tacit knowledge, or that they've started to think the way we think, or that they'll soon be able to do everything people can do. They don't, they haven't, and they won't. Artificial intelligence is not human intelligence. People are mindful; computers are mindless. But when it comes to performing demanding tasks, whether with the brain or the body, computers are able to replicate our ends without replicating our means. When a driverless car makes a left turn in traffic, it's not tapping into a well of intuition and skill; it's following a program. But while the strategies are different, the outcomes, for practical purposes, are the same. The superhuman speed with which computers can follow instructions, calculate probabilities, and receive and send data means that they can use explicit knowledge to perform many of the complicated tasks that we do with tacit knowledge. In some cases, the unique strengths of computers allow them to perform what we consider to be tacit skills better than we can perform them ourselves. In a world of computer-controlled cars, you wouldn't need traffic lights or stop signs. Through the continuous, high-speed exchange of data, vehicles would seamlessly coordinate their passage through even the busiest of intersections—just as computers today regulate the flow of inconceivable numbers of data packets along the highways and byways of the internet. What's ineffable in our own minds becomes altogether effable in the circuits of a microchip.

Many of the cognitive talents we've considered uniquely human, it turns out, are anything but. Once computers get quick enough, they can begin to mimic our ability to spot patterns, make judgments, and learn from experience. We were first taught that lesson back in 1997 when IBM's Deep Blue chess-playing supercomputer, which could evaluate a billion possible moves every five seconds, beat the world champion Garry Kasparov. With Google's intelligent car, which can process a million environmental readings a second, we're learning the lesson again. A lot of the very smart things that people do don't actually require a brain. The intellectual talents of highly trained professionals are no more protected from automation than is the driver's left turn. We see the evidence everywhere. Creative and analytical work of all sorts is being mediated by software. Doctors use computers to diagnose diseases. Architects use them to design buildings. Attorneys use them to evaluate evidence. Musicians use them to simulate instruments and correct bum notes. Teachers use them to tutor students and grade papers. Computers aren't taking over these professions entirely, but they are taking over many aspects of them. And they're certainly changing the way the work is performed.

It's not only vocations that are being computerized. Avocations are too. Thanks to the proliferation of smartphones, tablets, and other small, affordable, and even wearable computers, we now depend on software to carry out many of our daily chores and pastimes. We launch apps to aid us in shopping, cooking, exercising, even finding a mate and raising a child. We follow turn-by-turn GPS instructions to get from one place to the next. We use social networks to maintain friendships and express our feelings. We seek advice from recommendation engines on what to watch, read, and listen to. We look to Google, or to Apple's Siri, to answer our questions and solve our problems. The computer is becoming our all-purpose tool for navigating, manipulating, and understanding the world, in both its physical and its social manifestations. Just think what happens these days when people misplace their smartphones or lose their connections to the net. Without their digital assistants, they feel helpless. As Katherine Hayles, a literature professor at Duke University, observed in her 2012 book *How We Think*, "When my computer goes down or my Internet connection fails, I feel lost, disoriented, unable to work—in fact, I feel as if my hands have been amputated."[6]

Our dependency on computers may be disconcerting at times, but in general we welcome it. We're eager to celebrate and show off our whizzy new gadgets and apps—and not only because they're so useful and so stylish. There's something magical about computer automation. To watch an iPhone identify an obscure song playing over the sound system in a bar is to experience something that would have been inconceivable to any previous generation. To see a crew of brightly painted factory robots effortlessly assemble a solar panel or a jet engine is to view an exquisite heavy-metal ballet, each movement choreographed to a fraction of a millimeter and a sliver of a second. The people who have taken rides in Google's car report that the thrill is almost otherworldly; their earth-bound brain has a tough time processing the experience. Today, we really do seem to be entering a brave new world, a Tomorrowland where computers and automatons will be at our service, relieving us of our burdens, granting our wishes, and sometimes just keeping us company. Very soon now, our Silicon Valley wizards assure us, we'll have robot maids as well as robot chauffeurs. Sundries will be fabricated by 3-D printers and delivered to our doors by drones. The world of the *Jetsons*, or at least of *Knight Rider*, beckons.

It's hard not to feel awestruck. It's also hard not to feel apprehensive. An automatic transmission may seem a paltry thing beside Google's tricked-out, look-ma-no-humans Prius, but the former was a precursor to the latter, a small step along the path to total automation, and I can't help but remember the letdown I felt after the gear stick was taken from my hand—or, to put responsibility where it belongs, after I begged to have the gear stick taken from my hand. If the convenience of an automatic transmission left me feeling a little lacking, a little *underutilized*, as a labor economist might say, how will it feel to become, truly, a passenger in my own car?

■   ■   ■   ■

THE TROUBLE with automation is that it often gives us what we don't need at the cost of what we do. To understand why that's so, and why we're eager to accept the bargain, we need to take a look at how certain cognitive biases—flaws in the way we think—can distort our perceptions. When it comes to assessing the value of labor and leisure, the mind's eye can't see straight.

Mihaly Csikszentmihalyi, a psychology professor and author of the popular 1990 book *Flow*, has described a phenomenon that he calls "the paradox of work." He first observed it in a study he conducted in the 1980s with his University of Chicago colleague Judith LeFevre. They recruited a hundred workers, blue-collar and white-collar, skilled and unskilled, from five businesses around Chicago. They gave each an electronic pager (this was when cell phones were still luxury goods) that they had programmed to beep at seven random moments a day over the course of a week. At each beep, the subjects would fill out a short questionnaire. They'd describe the activity they were engaged in at that moment, the challenges they were facing, the skills they were deploying, and the psychological state they were in, as indicated by their sense of motivation, satisfaction, engagement, creativity, and so forth. The intent of this "experience sampling," as Csikszentmihalyi termed the technique, was to see how people spend their time, on the job and off, and how their activities influence their "quality of experience."

The results were surprising. People were happier, felt more fulfilled by what they were doing, while they were at work than during their leisure hours. In their free time, they tended to feel bored and anxious. And yet they didn't like to be at work. When they were on the job, they expressed a strong desire to be off the job, and when they were off the job, the last thing they wanted was to go back to work. "We have," reported Csikszentmihalyi and LeFevre, "the paradoxical situation of people having many more positive feelings at work than in leisure, yet saying that they 'wish to be doing something else' when they are at work, not when they are in leisure."[7] We're terrible, the experiment revealed, at anticipating which activities will satisfy us and which will leave us discontented. Even when we're in the midst of doing something, we don't seem able to judge its psychic consequences accurately.

Those are symptoms of a more general affliction, on which psychologists have bestowed the poetic name *miswanting*. We're inclined to desire things we don't like and to like things we don't desire. "When the things we want to happen do not improve our happiness, and when the things we want not to happen do," the cognitive psychologists Daniel Gilbert and Timothy Wilson have observed, "it seems fair to say we have wanted badly."[8] And as slews of gloomy studies show, we're forever wanting badly. There's also a social angle to our tendency to misjudge work and leisure. As Csikszentmihalyi and LeFevre discovered in their experiments, and as most of us know from our own experience, people allow themselves to be guided by social conventions—in this case, the deep-seated idea that being "at leisure" is more desirable, and carries more status, than being "at work"—rather than by their true feelings. "Needless to say," the researchers concluded, "such a blindness to the real state of affairs is likely to have unfortunate consequences for both individual well-being and the health of society." As people act on their skewed perceptions, they will "try to do more of those activities that provide the least positive experiences and avoid the activities that are the source of their most positive and intense feelings."[9] That's hardly a recipe for the good life.

It's not that the work we do for pay is intrinsically superior to the activities we engage in for diversion or entertainment. Far from it. Plenty of jobs are dull and even demeaning, and plenty of hobbies and pastimes are stimulating and fulfilling. But a job imposes a structure on our time that we lose when we're left to our own devices. At work, we're pushed to engage in the kinds of activities that human beings find most satisfying. We're happiest when we're absorbed in a difficult task, a task that has clear goals and that challenges us not only to exercise our talents but to stretch them. We become so immersed in the flow of our work, to use Csikszentmihalyi's term, that we tune out distractions and transcend the anxieties and worries that plague our everyday lives. Our usually wayward attention becomes fixed on what we're doing. "Every action, movement, and thought follows inevitably from the previous one," explains Csikszentmihalyi. "Your whole being is involved, and you're using your skills to the utmost."[10] Such states of deep absorption can be produced by all manner of effort, from laying tile to singing in a choir to racing a dirt bike. You don't have to be earning a wage to enjoy the transports of flow.

More often than not, though, our discipline flags and our mind wanders when we're not on the job. We may yearn for the workday to be over so we can start spending our pay and having some fun, but most of us fritter away our leisure hours. We shun hard work and only rarely engage in challenging hobbies. Instead, we watch TV or go to the mall or log on to Facebook. We get lazy. And then we get bored and fretful. Disengaged from any outward focus, our attention turns inward, and we end up locked in what Emerson called the jail of self-consciousness. Jobs, even crummy ones, are "actually easier to enjoy than free time," says Csikszentmihalyi, because they have the "built-in" goals and challenges that "encourage one to become involved in one's work, to concentrate and lose oneself in it."[11] But that's not what our deceiving minds want us to believe. Given the opportunity, we'll eagerly relieve ourselves of the rigors of labor. We'll sentence ourselves to idleness.

■   ■   ■   ■

Is it any wonder we're enamored of automation? By offering to reduce the amount of work we have to do, by promising to imbue our lives with greater ease, comfort, and convenience, computers and other labor-saving technologies appeal to our eager but misguided desire for release from what we perceive as toil. In the workplace, automation's focus on enhancing speed and efficiency—a focus determined by the profit motive rather than by any particular concern for people's well-being—often has the effect of removing complexity from jobs, diminishing the challenge they present and hence the engagement they promote. Automation can narrow people's responsibilities to the point that their jobs consist largely of monitoring a computer screen or entering data into prescribed fields. Even highly trained analysts and other so-called knowledge workers are seeing their work circumscribed by decision-support systems that turn the making of judgments into a data-processing routine. The apps and other programs we use in our private lives have similar effects. By taking over difficult or time-consuming tasks, or simply rendering those tasks less onerous, the software makes it even less likely that we'll engage in efforts that test our skills and give us a sense of accomplishment and satisfaction. All too often, automation frees us from that which makes us feel free.

The point is not that automation is bad. Automation and its precursor, mechanization, have been marching forward for centuries, and by and large our circumstances have improved greatly as a result. Deployed wisely, automation can relieve of us drudge work and spur us on to more challenging and fulfilling endeavors. The point is that we're not very good at thinking rationally about automation or understanding its implications. We don't know when to say "enough" or even "hold on a second." The deck is stacked, economically and emotionally, in automation's favor. The benefits of transferring work from people to machines and computers are easy to identify and measure. Businesses can run the numbers on capital investments and calculate automation's benefits in hard currency: reduced labor costs, improved productivity, faster throughputs and turnarounds, higher profits. In our personal lives, we can point to all sorts of ways that computers allow us to save time and avoid hassles. And thanks to our bias for leisure over work, ease over effort, we overestimate automation's benefits.

The costs are harder to pin down. We know computers make certain jobs obsolete and put some people out of work, but history suggests, and most economists assume, that any declines in employment will prove temporary and that over the long haul productivity-boosting technology will create attractive new occupations and raise standards of living. The personal costs are even hazier. How do you measure the expense of an erosion of effort and engagement, or a waning of agency and autonomy, or a subtle deterioration of skill? You can't. Those are the kinds of shadowy, intangible things that we rarely appreciate until after they're gone, and even then we may have trouble expressing the losses in concrete terms. But the costs are real. The choices we make, or fail to make, about which tasks we hand off to computers and which we keep for ourselves are not just practical or economic choices. They're ethical choices. They shape the substance of our lives and the place we make for ourselves in the world. Automation confronts us with the most important question of all: What does *human being* mean?

Csikszentmihalyi and LeFevre discovered something else in their study of people's daily routines. Among all the leisure activities reported by their test subjects, the one that generated the greatest sense of flow was driving a car.

CHAPTER TWO

THE ROBOT AT THE GATE

IN THE EARLY 1950S, LESLIE ILLINGWORTH, A MUCH-ADMIRED political cartoonist at the British satirical magazine *Punch*, drew a dark and foreboding sketch. Set at dusk on what appears to be a stormy autumn day, it shows a worker peering anxiously from the doorway of an anonymous manufacturing plant. One of his hands grips a small tool; the other is balled into a fist. He looks out across the muddy factory yard to the plant's main gate. There, standing beside a sign reading "Hands Wanted," looms a giant, broad-shouldered robot. Across its chest, emblazoned in block letters, is the word "Automation."

The illustration was a sign of its times, a reflection of a new anxiety seeping through Western society. In 1956, it was reprinted as the frontispiece of a slender but influential book called *Automation: Friend or Foe?* by Robert Hugh Macmillan, an engineering professor at Cambridge University. On the first page, Macmillan posed an unsettling question: "Are we in danger of being destroyed by our own creations?" He was not, he explained, referring to the well-known "perils of unrestricted 'push-button' warfare." He was talking about a less discussed but more insidious threat: "the rapidly increasing part that automatic devices are playing in the peace-time industrial life of all civilized countries."[1] Just as earlier machines "had replaced man's muscles," these new devices seemed likely to "replace his brains." By taking over many good, well-paying jobs, they threatened to create widespread unemployment, leading to social strife and upheaval—of just the sort Karl Marx had foreseen a century earlier.[2]

But, Macmillan continued, it didn't have to be that way. If *"rightly applied,"* automation could bring economic stability, spread prosperity, and relieve the human race of its toils. "My hope is that this new branch of technology may eventually enable us to lift the curse of Adam from the shoulders of man, for machines could indeed become men's slaves rather than their masters, now that practical techniques have been devised for controlling them automatically."[3] Whether technologies of automation ultimately proved boon or bane, Macmillan warned, one thing was certain: they would play an ever greater role in industry and society. The economic imperatives of "a highly competitive world" made that inevitable.[4] If a robot could work faster, cheaper, or better than its human counterpart, the robot would get the job.

■   ■   ■   ■

"WE ARE brothers and sisters of our machines," the technology historian George Dyson once remarked.[5] Sibling relations are notoriously fraught, and so it is with our technological kin. We love our machines—not just because they're useful to us, but because we find them companionable and even beautiful. In a well-built machine, we see some of our deepest aspirations take form: the desire to understand the world and its workings, the desire to turn nature's power to our own purposes, the desire to add something new and of our own fashioning to the cosmos, the desire to be awed and amazed. An ingenious machine is a source of wonder and of pride.

But machines are ugly too, and we sense in them a threat to things we hold dear. Machines may be a conduit of human power, but that power has usually been wielded by the industrialists and financiers who own the contraptions, not the people paid to operate them. Machines are cold and mindless, and in their obedience to scripted routines we see an image of society's darker possibilities. If machines bring something human to the alien cosmos, they also bring something alien to the human world. The mathematician and philosopher Bertrand Russell put it succinctly in a 1924 essay: "Machines are worshipped because they are beautiful and valued because they confer power; they are hated because they are hideous and loathed because they impose slavery."[6]

As Russell's comment suggests, the tension in Macmillan's view of automated machines—they'd either destroy us or redeem us, liberate us or enslave us—has a long history. The same tension has run through popular reactions to factory machinery since the start of the Industrial Revolution more than two centuries ago. While many of our forebears celebrated the arrival of mechanized production, seeing it as a symbol of progress and a guarantor of prosperity, others worried that machines would steal their jobs and even their souls. Ever since, the story of technology has been one of rapid, often disorienting change. Thanks to the ingenuity of our inventors and entrepreneurs, hardly a decade has passed without the arrival of new, more elaborate, and more capable machinery. Yet our ambivalence toward these fabulous creations, creations of our own hands and minds, has remained a constant. It's almost as if in looking at a machine we see, if only dimly, something about ourselves that we don't quite trust.

In his 1776 masterwork *The Wealth of Nations*, the foundational text of free enterprise, Adam Smith praised the great variety of "very pretty machines" that manufacturers were installing to "facilitate and abridge labour." By enabling "one man to do the work of many," he predicted, mechanization would provide a great boost to industrial productivity.[7] Factory owners would earn more profits, which they would then invest in expanding their operations—building more plants, buying more machines, hiring more employees. Each individual machine's abridgment of labor, far from being bad for workers, would actually stimulate demand for labor in the long run.

Other thinkers embraced and extended Smith's assessment. Thanks to the higher productivity made possible by labor-saving equipment, they predicted, jobs would multiply, wages would go up, and prices of goods would come down. Workers would have some extra cash in their pockets, which they would use to purchase products from the manufacturers that employed them. That would provide yet more capital for industrial expansion. In this way, mechanization would help set in motion a virtuous cycle, accelerating a society's economic growth, expanding and spreading its wealth, and bringing to its people what Smith had termed "convenience and luxury."[8] This view of technology as an economic elixir seemed, happily, to be borne out by the early history of industrialization, and it became a fixture of economic theory. The idea wasn't compelling only to early capitalists and their scholarly brethren. Many social reformers applauded mechanization, viewing it as the best hope for raising the urban masses out of poverty and servitude.

Economists, capitalists, and reformers could afford to take the long view. With the workers themselves, that wasn't the case. Even a temporary abridgment of labor could pose a real and immediate threat to their livelihoods. The installation of new factory machines put plenty of people out of jobs, and it forced others to exchange interesting, skilled work for the tedium of pulling levers and pressing foot-pedals. In many parts of Britain during the eighteenth and the early nineteenth century, skilled workers took to sabotaging the new machinery as a way to defend their jobs, their trades, and their communities. "Machine-breaking," as the movement came to be called, was not simply an attack on technological progress. It was a concerted attempt by tradesmen to protect their ways of life, which were very much bound up in the crafts they practiced, and to secure their economic and civic autonomy. "If the workmen disliked certain machines," writes the historian Malcolm Thomis, drawing on contemporary accounts of the uprisings, "it was because of the use to which they were being put, not because they were machines or because they were new."[9]

Machine-breaking culminated in the Luddite rebellion that raged through the industrial counties of the English Midlands from 1811 to 1816. Weavers and knitters, fearing the destruction of their small-scale, locally organized cottage industry, formed guerrilla bands with the intent of stopping big textile mills and factories from installing mechanized looms and stocking frames. The Luddites—the rebels took their now-notorious name from a legendary Leicestershire machine-breaker known as Ned Ludlam—launched nighttime raids against the plants, often wrecking the new equipment. Thousands of British troops had to be called in to battle the rebels, and the soldiers put down the revolt with brutal force, killing many and incarcerating others.

Although the Luddites and other machine-breakers had some scattered success in slowing the pace of mechanization, they certainly didn't stop it. Machines were soon so commonplace in factories, so essential to industrial production and competition, that resisting their use came to be seen as an exercise in futility. Workers acquiesced to the new technological regime, though their distrust of machinery persisted.

■   ■   ■   ■

IT WAS Marx who, a few decades after the Luddites lost their fight, gave the deep divide in society's view of mechanization its most powerful and influential expression. Frequently in his writings, Marx invests factory machinery with a demonic, parasitic will, portraying it as "dead labour" that "dominates, and pumps dry, living labour power." The workman becomes a "mere living appendage" of the "lifeless mechanism."[10] In a darkly prophetic remark during an 1856 speech, he said, "All our invention and progress seem to result in endowing material forces with intellectual life, and stultifying human life into a material force."[11] But Marx didn't just talk about the "infernal effects" of machines. As the media scholar Nick Dyer-Witheford has explained, he also saw and lauded "their emancipatory promise."[12] Modern machinery, Marx observed in that same speech, has "the wonderful power of shortening and fructifying human labour."[13] By freeing workers from the narrow specializations of their trades, machines might allow them to fulfill their potential as "totally developed" individuals, able to shift between "different modes of activity" and hence "different social functions."[14] In the right hands—those of the workers rather than the capitalists—technology would no longer be the yoke of oppression. It would become the uplifting block and tackle of self-fulfillment.

The idea of machines as emancipators took stronger hold in Western culture as the twentieth century approached. In an 1897 article praising the mechanization of American industry, the French economist Émile Levasseur ticked off the benefits that new technology had brought to "the laboring classes." It had raised workers' wages and pushed down the prices they paid for goods, providing them with greater material comfort. It had spurred a redesign of factories, making workplaces cleaner, better lit, and generally more hospitable than the dark satanic mills that characterized the early years of the Industrial Revolution. Most important of all, it had elevated the kind of work that factory hands performed. "Their task has become less onerous, the machine doing everything which requires great strength; the workman, instead of bringing his muscles into play, has become an inspector, using his intelligence." Levasseur acknowledged that laborers still grumbled about having to operate machinery. "They reproach [the machine] with demanding such continued attention that it enervates," he wrote, and they accuse it of "degrading man by transforming him into a machine, which knows how to make but one movement, and that always the same." Yet he dismissed such complaints as blinkered. The workers simply didn't understand how good they had it.[15]

Some artists and intellectuals, believing the imaginative work of the mind to be inherently superior to the productive labor of the body, saw a technological utopia in the making. Oscar Wilde, in an essay published at about the same time as Levasseur's, though aimed at a very different audience, foresaw a day when machines would not just alleviate toil but eliminate it. "All unintellectual labour, all monotonous, dull labour, all labour that deals with dreadful things, and involves unpleasant conditions, must be done by machinery," he wrote. "On mechanical slavery, on the slavery of the machine, the future of the world depends." That machines would assume the role of slaves seemed to Wilde a foregone conclusion: "There is no doubt at all that this is the future of machinery, and just as trees grow while the country gentleman is asleep, so while Humanity will be amusing itself, or enjoying cultivated leisure—which, and not labour, is the aim of man—or making beautiful things, or reading beautiful things, or simply contemplating the world with admiration and delight, machinery will be doing all the necessary and unpleasant work."[16]

The Great Depression of the 1930s curbed such enthusiasm. The economic collapse prompted a bitter outcry against what had, in the Roaring Twenties, come to be known and celebrated as the Machine Age. Labor unions and religious groups, crusading editorial writers and despairing citizens—all railed against the job-destroying machines and the greedy businessmen who owned them. "Machinery did not inaugurate the phenomenon of unemployment," wrote the author of a best-selling book called *Men and Machines*, "but promoted it from a minor irritation to one of the chief plagues of mankind." It appeared, he went on, that "from now on, the better able we are to produce, the worse we shall be off."[17] The mayor of Palo Alto, California, wrote a letter to President Herbert Hoover imploring him to take action against the "Frankenstein monster" of industrial technology, a scourge that was "devouring our civilization."[18] At times the government itself inflamed the public's fears. One report issued by a federal agency called the factory machine "as dangerous as a wild animal." The uncontrolled acceleration of progress, its author wrote, had left society chronically unprepared to deal with the consequences.[19]

But the Depression did not entirely extinguish the Wildean dream of a machine paradise. In some ways, it rendered the utopian vision of progress more vivid, more necessary. The more we saw machines as our foes, the more we yearned for them to be our friends. "We are being afflicted," wrote the great British economist John Maynard Keynes in 1930, "with a new disease of which some readers may not yet have heard the name, but of which they will hear a great deal in the years to come—namely, *technological unemployment*." The ability of machines to take over jobs had outpaced the economy's ability to create valuable new work for people to do. But the problem, Keynes assured his readers, was merely a symptom of "a temporary phase of maladjustment." Growth and prosperity would return. Per-capita income would rise. And soon, thanks to the ingenuity and efficiency of our mechanical slaves, we wouldn't have to worry about jobs at all. Keynes thought it entirely possible that in a hundred years, by the year 2030, technological progress would have freed humankind from "the struggle for subsistence" and propelled us to "our destination of economic bliss." Machines would be doing even more of our work for us, but that would no longer be cause for worry or despair. By then, we would have figured out how to spread material wealth to everyone. Our only problem would be to figure out how to put our endless hours of leisure to good use—to teach ourselves "to enjoy" rather than "to strive."[20]

We're still striving, and it seems a safe bet that economic bliss will not have descended upon the planet by 2030. But if Keynes let his hopes get the best of him in the dark days of 1930, he was fundamentally right about the economy's prospects. The Depression did prove temporary. Growth returned, jobs came back, incomes shot up, and companies continued buying more and better machines. Economic equilibrium, imperfect and fragile as always, reestablished itself. Adam Smith's virtuous cycle kept turning.

By 1962, President John F. Kennedy could proclaim, during a speech in West Virginia, "We believe that if men have the talent to invent new machines that put men out of work, they have the talent to put those men back to work."[21] From the opening "we believe," the sentence is ringingly Kennedyesque. The simple words become resonant as they're repeated: *men, talent, men, work, talent, men, work.* The drum-like rhythm marches forward, giving the stirring conclusion—"back to work"—an air of inevitability. To those listening, Kennedy's words must have sounded like the end of the story. But they weren't. They were the end of one chapter, and a new chapter had already begun.

■   ■   ■   ■

WORRIES ABOUT technological unemployment have been on the rise again, particularly in the United States. The recession of the early 1990s, which saw exalted U.S. companies such as General Motors, IBM, and Boeing fire tens of thousands of workers in massive "restructurings," prompted fears that new technologies, particularly cheap computers and clever software, were about to wipe out middle-class jobs. In 1994, the sociologists Stanley Aronowitz and William DiFazio published *The Jobless Future*, a book that implicated "labor-displacing technological change" in "the trend toward more low-paid, temporary, benefit-free blue- and white-collar jobs and fewer decent *permanent* factory and office jobs."[22] The following year, Jeremy Rifkin's unsettling *The End of Work* appeared. The rise of computer automation had inaugurated a "Third Industrial Revolution," declared Rifkin. "In the years ahead, new, more sophisticated software technologies are going to bring civilization ever closer to a near workerless world." Society had reached a turning point, he wrote. Computers could "result in massive unemployment and a potential global depression," but they could also "free us for a life of increasing leisure" if we were willing to rewrite the tenets of contemporary capitalism.[23] The two books, and others like them, caused a stir, but once again fears about technology-induced joblessness passed quickly. The resurgence of economic growth through the middle and late 1990s, culminating in the giddy dot-com boom, turned people's attention away from apocalyptic predictions of mass unemployment.

A decade later, in the wake of the Great Recession of 2008, the anxieties returned, stronger than ever. In mid-2009, the American economy, recovering fitfully from the economic collapse, began to expand again. Corporate profits rebounded. Businesses ratcheted their capital investments up to pre-recession levels. The stock market soared. But hiring refused to bounce back. While it's not unusual for companies to wait until a recovery is well established before recruiting new workers, this time the hiring lag seemed interminable. Job growth remained unusually tepid, the unemployment rate stubbornly high. Seeking an explanation, and a culprit, people looked to the usual suspect: labor-saving technology.

Late in 2011, two respected MIT researchers, Erik Brynjolfsson and Andrew McAfee, published a short electronic book, *Race against the Machine*, in which they gently chided economists and policy makers for dismissing the possibility that workplace technology was substantially reducing companies' need for new employees. The "empirical fact" that machines had bolstered employment for centuries "conceals a dirty secret," they wrote. "There is no economic law that says that everyone, or even most people, automatically benefit from technological progress." Although Brynjolfsson and McAfee were anything but technophobes—they remained "hugely optimistic" about the ability of computers and robots to boost productivity and improve people's lives over the long run—they made a strong case that technological unemployment was real, that it had become pervasive, and that it would likely get much worse. Human beings, they warned, were losing the race against the machine.[24]

Their ebook was like a match thrown onto a dry field. It sparked a vigorous and sometimes caustic debate among economists, a debate that soon drew the attention of journalists. The phrase "technological unemployment," which had faded from use after the Great Depression, took a new grip on the public mind. At the start of 2013, the TV news program *60 Minutes* ran a segment, called "March of the Machines," that examined how businesses were using new technologies in place of workers at warehouses, hospitals, law firms, and manufacturing plants. Correspondent Steve Kroft lamented "a massive high-tech industry that's contributed enormous productivity and wealth to the American economy but surprisingly little in the way of employment."[25] Shortly after the program aired, a team of Associated Press writers published a three-part investigative report on the persistence of high unemployment. Their grim conclusion: jobs are "being obliterated by technology." Noting that science-fiction writers have long "warned of a future when we would be architects of our own obsolescence, replaced by our machines," the AP reporters declared that "the future has arrived."[26] They quoted one analyst who predicted that the unemployment rate would reach 75 percent by the century's end.[27]

Such forecasts are easy to dismiss. Their alarmist tone echoes the refrain heard time and again since the eighteenth century. Out of every economic downturn rises the specter of a job-munching Frankenstein monster. And then, when the economic cycle emerges from its trough and jobs return, the monster goes back in its cage and the worries subside. This time, though, the economy isn't behaving as it normally does. Mounting evidence suggests that a troubling new dynamic may be at work. Joining Brynjolfsson and McAfee, several prominent economists have begun questioning their profession's cherished assumption that technology-fueled productivity gains will bring job and wage growth. They point out that over the last decade U.S. productivity rose at a faster pace than we saw in the preceding thirty years, that corporate profits have hit levels we haven't seen in half a century, and that business investments in new equipment have been rising sharply. That combination should bring robust employment growth. And yet the total number of jobs in the country has barely budged. Growth and employment are "diverging in advanced countries," says economist Michael Spence, a Nobel laureate, and technology is the main reason why: "The replacement of routine manual jobs by machines and robots is a powerful, continuing, and perhaps accelerating trend in manufacturing and logistics, while networks of computers are replacing routine white-collar jobs in information processing."[28]

Some of the heavy spending on robots and other automation technologies in recent years may reflect temporary economic conditions, particularly the ongoing efforts by politicians and central banks to stimulate growth. Low interest rates and aggressive government tax incentives for capital investment have likely encouraged companies to buy labor-saving equipment and software that they might not otherwise have purchased.[29] But deeper and more prolonged trends also seem to be at work. Alan Krueger, the Princeton economist who chaired Barack Obama's Council of Economic Advisers from 2011 to 2013, points out that even before the recession "the U.S. economy was not creating enough jobs, particularly not enough middle-class jobs, and we were losing manufacturing jobs at an alarming rate."[30] Since then, the picture has only darkened. It might be assumed that, at least when it comes to manufacturing, jobs aren't disappearing but simply migrating to countries with low wages. That's not so. The total number of worldwide manufacturing jobs has been falling for years, even in industrial powerhouses like China, while overall manufacturing output has grown sharply.[31] Machines are replacing factory workers faster than economic expansion creates new manufacturing positions. As industrial robots become cheaper and more adept, the gap between lost and added jobs will almost certainly widen. Even the news that companies like GE and Apple are bringing some manufacturing work back to the United States is bittersweet. One of the reasons the work is returning is that most of it can be done without human beings. "Factory floors these days are nearly empty of people because software-driven machines are doing most of the work," reports economics professor Tyler Cowen.[32] A company doesn't have to worry about labor costs if it's not employing laborers.

The industrial economy—the economy of machines—is a recent phenomenon. It has been around for just two and a half centuries, a tick of history's second hand. Drawing definitive conclusions about the link between technology and employment from such limited experience was probably rash. The logic of capitalism, when combined with the history of scientific and technological progress, would seem to be a recipe for the eventual removal of labor from the processes of production. Machines, unlike workers, don't demand a share of the returns on capitalists' investments. They don't get sick or expect paid vacations or demand yearly raises. For the capitalist, labor is a problem that progress solves. Far from being irrational, the fear that technology will erode employment is fated to come true "in the very long run," argues the eminent economic historian Robert Skidelsky: "Sooner or later, we will run out of jobs."[33]

How long is the very long run? We don't know, though Skidelsky warns that it may be "uncomfortably close" for some countries.[34] In the near term, the impact of modern technology may be felt more in the distribution of jobs than in the overall employment figures. The mechanization of manual labor during the Industrial Revolution destroyed some good jobs, but it led to the creation of vast new categories of middle-class occupations. As companies expanded to serve bigger and more far-flung markets, they hired squads of supervisors and accountants, designers and marketers. Demand grew for teachers, doctors, lawyers, librarians, pilots, and all sorts of other professionals. The makeup of the job market is never static; it changes in response to technological and social trends. But there's no guarantee that the changes will always benefit workers or expand the middle class. With computers being programmed to take over white-collar work, many professionals are being forced into lower-paying jobs or made to trade full-time posts for part-time ones.

While most of the jobs lost during the recent recession were in well-paying industries, nearly three-fourths of the jobs created since the recession are in low-paying sectors. Having studied the causes of the "incredibly anemic employment growth" in the United States since 2000, MIT economist David Autor concludes that information technology "has really changed the distribution of occupation," creating a widening disparity in incomes and wealth. "There is an abundance of work to do in food service and there is an abundance of work in finance, but there are fewer middle-wage, middle-income jobs."[35] As new computer technologies extend automation into even more branches of the economy, we're likely to see an acceleration of this trend, with a further hollowing of the middle class and a growing loss of jobs among even the highest-paid professionals. "Smart machines may make higher GDP possible," notes Paul Krugman, another Nobel Prize–winning economist, "but also reduce the demand for people—including smart people. So we could be look-ing at a society that grows ever richer, but in which all the gains in wealth accrue to whoever owns the robots."[36]

The news is not all dire. As the U.S. economy gained steam dur-ing the second half of 2013, hiring strengthened in several sectors, including construction and health care, and there were encouraging gains in some higher-paying professions. The demand for workers remains tied to the economic cycle, if not quite so tightly as in the past. The increasing use of computers and software has itself created some very attractive new jobs as well as plenty of entrepreneurial opportunities. By historical standards, though, the number of people employed in computing and related fields remains modest. We can't all become software programmers or robotics engineers. We can't all decamp to Silicon Valley and make a killing writing nifty smart-phone apps.* With average wages stagnant and corporate profits con-tinuing to surge, the economy's bounties seem likely to go on flowing to the lucky few. And JFK's reassuring words will sound more and more suspect.

Why might this time be different? What exactly has changed that may be severing the old link between new technologies and new jobs? To answer that question we have to look back to that giant robot standing at the gate in Leslie Illingworth's cartoon—the robot named Automation.

---

* The internet, it's often noted, has opened opportunities for people to make money through their own personal initiative, with little investment of capital. They can sell used goods through eBay or crafts through Etsy. They can rent out a spare room through Airbnb or turn their car into a ghost cab with Lyft. They can find odd jobs through TaskRabbit. But while it's easy to pick up spare change through such modest enterprise, few people are going to be able to earn a middle-class income from the work. The real money goes to the software com-panies running the online clearinghouses that connect buyer and seller or lessor and lessee—clearinghouses that, being highly automated themselves, need few employees.

■   ■   ■   ■

THE WORD *automation* entered the language fairly recently. As best we can tell, it was first spoken in 1946, when engineers at the Ford Motor Company felt the need to coin a term to describe the latest machinery being installed on the company's assembly lines. "Give us some more of that automatic business," a Ford vice president reportedly said in a meeting. "Some more of that—that—'automation.'"[37] Ford's plants were already famously mechanized, with sophisticated machines streamlining every job on the line. But factory hands still had to lug parts and subassemblies from one machine to the next. The workers still controlled the pace of production. The equipment installed in 1946 changed that. Machines took over the material-handling and conveyance functions, allowing the entire assembly process to proceed automatically. The alteration in work flow may not have seemed momentous to those on the factory floor. But it was. Control over a complex industrial process had shifted from worker to machine.

The new word spread quickly. Two years later, in a report on the Ford machinery, a writer for the magazine *American Machinist* defined automation as "the art of applying mechanical devices to manipulate work pieces . . . in timed sequence with the production equipment so that the line can be put wholly or partially under push-button control at strategic stations."[38] As automation reached into more industries and production processes, and as it began to take on metaphorical weight in the culture, its definition grew more diffuse. "Few words of recent years have been so twisted to suit a multitude of purposes and phobias as this new word, 'automation,'" grumbled a Harvard business professor in 1958. "It has been used as a techno-logical rallying cry, a manufacturing goal, an engineering challenge, an advertising slogan, a labor campaign banner, and as the symbol of ominous technological progress." He then offered his own, eminently pragmatic definition: "Automation simply means something *significantly more automatic than previously existed in that plant, industry, or location.*"[39] Automation wasn't a thing or a technique so much as a force. It was more a manifestation of progress than a particular mode of operation. Any attempt at explaining or predicting its consequences would necessarily be tentative. As with many technological trends, automation would always be both old and new, and it would require a fresh reevaluation at each stage of its advance.

That Ford's automated equipment arrived just after the end of the Second World War was no accident. It was during the war that modern automation technology took shape. When the Nazis began their bombing blitz against Great Britain in 1940, English and American scientists faced a challenge as daunting as it was pressing: How do you knock high-flying, fast-moving bombers out of the sky with heavy missiles fired from unwieldy antiaircraft guns on the ground? The mental calculations and physical adjustments required to aim a gun accurately—not at a plane's current position but at its probable future position—were far too complicated for a soldier to perform with the speed necessary to get a shot off while a plane was still in range. This was no job for mortals. The missile's trajectory, the scientists saw, had to be computed by a calculating machine, using tracking data coming in from radar systems along with statistical projections of a plane's course, and then the calculations had to be fed automatically into the gun's aiming mechanism to guide the firing. The gun's aim, moreover, had to be adjusted continually to account for the success or failure of previous shots.

As for the members of the gunnery crews, their work would have to change to accommodate the new generation of automated weapons. And change it did. Artillerymen soon found themselves sitting in front of screens in darkened trucks, selecting targets from radar displays. Their identities shifted along with their jobs. They were no longer seen "as soldiers," writes one historian, but rather "as technicians reading and manipulating representations of the world."[40]

In the antiaircraft cannons born of the Allied scientists' work, we see all the elements of what now characterizes an automated system. First, at the system's core, is a very fast calculating machine—a computer. Second is a sensing mechanism (radar, in this case) that monitors the external environment, the real world, and communicates essential data about it to the computer. Third is a communication link that allows the computer to control the movements of the physical apparatus that performs the actual work, with or without human assistance. And finally there's a feedback method—a means of returning to the computer information about the results of its instructions so that it can adjust its calculations to correct for errors and account for changes in the environment. Sensory organs, a calculating brain, a stream of messages to control physical movements, and a feedback loop for learning: there you have the essence of automation, the essence of a robot. And there, too, you have the essence of a living being's nervous system. The resemblance is no coincidence. In order to replace a human, an automated system first has to replicate a human, or at least some aspect of a human's ability.

Automated machines existed before World War II. James Watt's steam engine, the original prime mover of the Industrial Revolution, incorporated an ingenious feedback device—the fly-ball governor—that enabled it to regulate its own operation. As the engine sped up, it rotated a pair of metal balls, creating a centrifugal force that pulled a lever to close a steam valve, keeping the engine from running too fast. The Jacquard loom, invented in France around 1800, used steel punch cards to control the movements of spools of different-colored threads, allowing intricate patterns to be woven automatically. In 1866, a British engineer named J. Macfarlane Gray patented a steam-ship steering mechanism that was able to register the movement of a boat's helm and, through a gear-operated feedback system, adjust the angle of the rudder to maintain a set course.[41] But the development of fast computers, along with other sensitive electronic controls, opened a new chapter in the history of machines. It vastly expanded the possibilities of automation. As the mathematician Norbert Wiener, who helped write the prediction algorithms for the Allies' automated anti-aircraft gun, explained in his 1950 book *The Human Use of Human Beings*, the advances of the 1940s enabled inventors and engineers to go beyond "the sporadic design of individual automatic mechanisms." The new technologies, while designed with weaponry in mind, gave rise to "a general policy for the construction of automatic mechanisms of the most varied type." They paved the way for "the new automatic age."[42]

Beyond the pursuit of progress and productivity lay another impetus for the automatic age: politics. The postwar years were characterized by intense labor strife. Managers and unions battled in most American manufacturing sectors, and the tensions were often strongest in industries essential to the federal government's Cold War buildup of military equipment and armaments. Strikes, walkouts, and slowdowns were daily events. In 1950 alone, eighty-eight work stoppages were staged at a single Westinghouse plant in Pittsburgh. In many factories, union stewards held more power over operations than did corporate managers—the workers called the shots. Military and industrial planners saw automation as a way to shift the balance of power back to management. Electronically controlled machinery, declared *Fortune* magazine in a 1946 cover story titled "Machines without Men," would prove "immensely superior to the human mechanism," not least because machines "are always satisfied with working conditions and never demand higher wages."[43] An executive with Arthur D. Little, a leading management and engineering consultancy, wrote that the rise of automation heralded the business world's "emancipation from human workers."[44]

In addition to reducing the need for laborers, particularly skilled ones, automated equipment provided business owners and managers with a technological means to control the speed and flow of production through the electronic programming of individual machines and entire assembly lines. When, at the Ford plants, control over the pace of the line shifted to the new automated equipment, the workers lost a great deal of autonomy. By the mid-1950s, the role of labor unions in charting factory operations was much diminished.[45] The lesson would prove important: in an automated system, power concentrates with those who control the programming.

Wiener foresaw, with uncanny clarity, what would come next. The technologies of automation would advance far more rapidly than anyone had imagined. Computers would get faster and smaller. The speed and capacity of electronic communication and storage systems would increase exponentially. Sensors would see, hear, and feel the world with ever greater sensitivity. Robotic mechanisms would come "to replicate more nearly the functions of the human hand as supplemented by the human eye." The cost to manufacture all the new devices and systems would plummet. The use of automation would become both possible and economical in ever more areas. And since computers could be programmed to carry out logical functions, automation's reach would extend beyond the work of the hand and into the work of the mind—the realm of analysis, judgment, and decision making. A computerized machine didn't have to act by manipulating material things like guns. It could act by manipulating information. "From this stage on, everything may go by machine," Wiener wrote. "The machine plays no favorites between manual labor and white-collar labor." It seemed obvious to him that automation would, sooner or later, create "an unemployment situation" that would make the calamity of the Great Depression "seem a pleasant joke."[46]

The Human Use of Human Beings was a best seller, as was Wiener's earlier and much more technical treatise, *Cybernetics, or Control and Communication in the Animal and the Machine*. The mathematician's unsettling analysis of technology's trajectory became part of the intellectual texture of the 1950s. It inspired or informed many of the books and articles on automation that appeared during the decade, including Robert Hugh Macmillan's slim volume. An aging Bertrand Russell, in a 1951 essay, "Are Human Beings Necessary?," wrote that Wiener's work made it clear that "we shall have to change some of the fundamental assumptions upon which the world has been run ever since civilization began."[47] Wiener even makes a brief appearance as a forgotten prophet in Kurt Vonnegut's first novel, the 1952 dystopian satire *Player Piano*, in which a young engineer's rebellion against a rigidly automated world ends with an epic episode of machine-breaking.

＊　　＊　　＊　　＊

THE IDEA of a robot invasion may have seemed threatening, if not apocalyptic, to a public already rattled by the bomb, but automation technologies were still in their infancy during the 1950s. Their ultimate consequences could be imagined, in speculative tracts and science-fiction fantasies, but those consequences were still a long way from being experienced. Through the 1960s, most automated machines continued to resemble the primitive robotic haulers on Ford's postwar assembly lines. They were big, expensive, and none too bright. Most of them could perform only a single, repetitive function, adjusting their movements in response to a few elementary electronic commands: speed up, slow down; move left, move right; grasp, release. The machines were extraordinarily precise, but otherwise their talents were few. Toiling anonymously inside factories, often locked within cages to protect passersby from their mindless twists and jerks, they certainly didn't look like they were about to take over the world. They seemed little more than very well-behaved and well-coordinated beasts of burden.

But robots and other automated systems had one big advantage over the purely mechanical contraptions that came before them. Because they ran on software, they could hitch a ride on the Moore's Law Express. They could benefit from all the rapid advances—in processor speed, programming algorithms, storage and network capacity, interface design, and miniaturization—that came to characterize the progress of computers themselves. And that, as Wiener predicted, is what happened. Robots' senses grew sharper; their brains, quicker and more supple; their conversations, more fluent; their ability to learn, more capacious. By the early 1970s, they were taking over production work that required flexibility and dexterity—cutting, welding, assembling. By the end of that decade, they were flying planes as well as building them. And then, freed from their physical embodiments and turned into the pure logic of code, they spread out into the business world through a multitude of specialized software applications. They entered the cerebral trades of the white-collar workforce, sometimes as replacements but far more often as assistants.

Robots may have been at the factory gate in the 1950s, but it's only recently that they've marched, on our orders, into offices, shops, and homes. Today, as software of what Wiener termed "the judgment-replacing type" moves from our desks to our pockets, we're at last beginning to experience automation's true potential for changing what we do and how we do it. Everything is being automated. Or, as Netscape founder and Silicon Valley grandee Marc Andreessen puts it, "software is eating the world."[48]

That may be the most important lesson to be gleaned from Wiener's work—and, for that matter, from the long, tumultuous history of labor-saving machinery. Technology changes, and it changes more quickly than human beings change. Where computers sprint forward at the pace of Moore's law, our own innate abilities creep ahead with the tortoise-like tread of Darwin's law. Where robots can be constructed in a myriad of forms, replicating everything from snakes that burrow in the ground to raptors that swoop across the sky to fish that swim through the sea, we're basically stuck with our old, forked bodies. That doesn't mean our machines are about to leave us in the evolutionary dust. Even the most powerful supercomputer evidences no more consciousness than a hammer. It does mean that our software and our robots will, with our guidance, continue to find new ways to outperform us—to work faster, cheaper, better. And, like those antiaircraft gunners during World War II, we'll be compelled to adapt our own work, behavior, and skills to the capabilities and routines of the machines we depend on.

CHAPTER THREE

ON AUTOPILOT

On the evening of February 12, 2009, a Continental Connection commuter flight made its way through blustery weather between Newark, New Jersey, and Buffalo, New York. As is typical of commercial flights these days, the two pilots didn't have all that much to do during the hour-long trip. The captain, an affable, forty-seven-year-old Floridian named Marvin Renslow, manned the controls briefly during takeoff, guiding the Bombardier Q400 turboprop into the air, then switched on the autopilot. He and his cabin mate, twenty-four-year-old first officer Rebecca Shaw, a newlywed from Seattle, kept an eye on the computer readouts that flickered across the cockpit's five large LCD screens. They exchanged some messages over the radio with air traffic controllers. They went through a few routine checklists. Mostly, though, they passed the time chatting amiably about this and that—families, careers, colleagues, money—as the turboprop cruised along its northwesterly route at sixteen thousand feet.[1]

The Q400 was well into its approach to the Buffalo airport, its landing gear down, its wing flaps out, when the captain's control yoke began to shudder noisily. The plane's "stick shaker" had activated, a signal that the turboprop was losing lift and risked going into an aero-dynamic stall.* The autopilot disconnected, as it's programmed to do in the event of a stall warning, and the captain took over the con-trols. He reacted quickly, but he did precisely the wrong thing. He jerked back on the yoke, lifting the plane's nose and reducing its air speed, instead of pushing the yoke forward to tip the craft down and gain velocity. The plane's automatic stall-avoidance system kicked in and attempted to push the yoke forward, but the captain simply redoubled his effort to pull it back toward him. Rather than prevent a stall, Renslow caused one. The Q400 spun out of control, then plummeted. "We're down," the captain said, just before the plane slammed into a house in a Buffalo suburb.

The crash, which killed all forty-nine people onboard as well as one person on the ground, should not have happened. A National Transportation Safety Board investigation found no evidence of mechanical problems with the Q400. Some ice had accumulated on the plane, but nothing out of the ordinary for a winter flight. The deicing equipment had operated properly, as had the plane's other systems. Renslow had had a fairly demanding flight schedule over the preceding two days, and Shaw had been battling a cold, but both pilots seemed lucid and wakeful while in the cockpit. They were well trained, and though the stick shaker took them by surprise, they had plenty of time and airspace to make the adjustments necessary to avoid a stall. The NTSB concluded that the cause of the accident was pilot error. Neither Renslow nor Shaw had detected "explicit cues" that a stall warning was imminent, an oversight that suggested "a significant breakdown in their monitoring responsibilities." Once the warning sounded, the investigators reported, the captain's response

---

* A note on terminology: When people talk about a stall, they're usually referring to a loss of power in an engine. In aviation, a stall refers to a loss of lift in a wing.

"should have been automatic, but his improper flight control inputs were inconsistent with his training" and instead revealed "startle and confusion." An executive from the company that operated the flight for Continental, the regional carrier Colgan Air, admitted that the pilots seemed to lack "situational awareness" as the emergency unfolded.[2] Had the crew acted appropriately, the plane would likely have landed safely.

The Buffalo crash was not an isolated incident. An eerily similar disaster, with far more casualties, occurred a few months later. On the night of May 31, an Air France Airbus A330 took off from Rio de Janeiro, bound for Paris.[3] The jet ran into a storm over the Atlantic about three hours after takeoff. Its air-speed sensors, caked with ice, began giving faulty readings, which caused the autopilot to disengage. Bewildered, the copilot flying the plane, Pierre-Cédric Bonin, yanked back on the control stick. The A330 rose and a loud stall warning sounded, but Bonin continued to pull back heedlessly on the stick. As the plane climbed sharply, it lost velocity. The air-speed sensors began working again, providing the crew with accurate numbers. It should have been clear at this point that the jet was going too slow. Yet Bonin persisted in his mistake at the controls, causing a further deceleration. The jet stalled and began to fall. If Bonin had simply let go of the stick, the A330 might well have righted itself. But he didn't. The flight crew was suffering what French investigators would later term a "total loss of cognitive control of the situation."[4] After a few more harrowing seconds, another pilot, David Robert, took over the controls. It was too late. The plane dropped more than thirty thousand feet in three minutes.

"This can't be happening," said Robert.

"But what *is* happening?" replied the still-bewildered Bonin.

Three seconds later, the jet hit the ocean. All 228 crew and passengers died.

■   ■   ■   ■

IF YOU want to understand the human consequences of automation, the first place to look is up. Airlines and plane manufacturers, as well as government and military aviation agencies, have been particularly aggressive and especially ingenious in finding ways to shift work from people to machines. What car designers are doing with computers today, aircraft designers did decades ago. And because a single mistake in a cockpit can cost scores of lives and many millions of dollars, a great deal of private and public money has gone into funding psychological and behavioral research on automation's effects. For decades, scientists and engineers have been studying the ways automation influences the skills, perceptions, thoughts, and actions of pilots. Much of what we know about what happens when people work in concert with computers comes out of this research.

The story of flight automation begins a hundred years ago, on June 18, 1914, in Paris. The day was, by all accounts, a sunny and pleasant one, the blue sky a perfect backdrop for spectacle. A large crowd had gathered along the banks of the Seine, near the Argenteuil bridge in the city's northwestern fringes, to witness the Concours de la Sécurité en Aéroplane, an aviation competition organized to show off the latest advances in flight safety.[5] Nearly sixty planes and pilots took part, demonstrating an impressive assortment of techniques and equipment. Last on the day's program, flying a Curtiss C-2 biplane, was a handsome American pilot named Lawrence Sperry. Sitting beside him in the C-2's open cockpit was his French mechanic, Emil Cachin. As Sperry flew past the ranks of spectators and approached the judges' stand, he let go of the plane's controls and raised his hands. The crowd roared. The plane was flying itself!

Sperry was just getting started. After swinging the plane around, he took another pass by the reviewing stand, again with his hands in the air. This time, though, he had Cachin climb out of the cockpit and walk along the lower right wing, holding the struts between the wings for support. The plane tilted starboard for a second under the Frenchman's weight, then immediately righted itself, with no help from Sperry. The crowd roared even louder. Sperry circled around once again. By the time his plane approached the stands for its third pass, not only was Cachin out on the right wing, but Sperry himself had climbed out onto the left wing. The C-2 was flying, steady and true, with no one in the cockpit. The crowd and the judges were dumbfounded. Sperry won the grand prize—fifty thousand francs— and the next day his face beamed from the front pages of newspapers across Europe.

Inside the Curtiss C-2 was the world's first automatic pilot. Known as a "gyroscopic stabilizer apparatus," the device had been invented two years earlier by Sperry and his father, the famed American engineer and industrialist Elmer A. Sperry. It consisted of a pair of gyroscopes, one mounted horizontally, the other vertically, installed beneath the pilot's seat and powered by a wind-driven generator behind the propeller. Spinning at thousands of revolutions a minute, the gyroscopes were able to sense, with remarkable precision, a plane's orientation along its three axes of rotation—its lateral pitch, longitudinal roll, and vertical yaw. Whenever the plane diverged from its intended attitude, charged metal brushes attached to the gyroscopes would touch contact points on the craft's frame, completing a circuit. An electric current would flow to the motors operating the plane's main control panels—the ailerons on the wings and the elevators and rudder on the tail—and the panels would automatically adjust their positions to correct the problem. The horizontal gyroscope kept the plane's wings steady and its keel even, while the vertical one handled the steering.

It took nearly twenty years of further testing and refinement, much of it carried out under the auspices of the U.S. military, before the gyroscopic autopilot was ready to make its debut in commercial flight. But when it did, the technology still seemed as miraculous as ever. In 1930, a writer from *Popular Science* gave a breathless account of how an autopilot-equipped plane—"a big tri-motored Ford"—flew "without human aid" during a three-hour trip from Dayton, Ohio, to Washington, D.C. "Four men leaned back at ease in the passenger cabin," the reporter wrote. "Yet the pilot's compartment was empty. A metal airman, scarcely larger than an automobile battery, was holding the stick."[6] When, three years later, the daring American pilot Wiley Post completed the first solo flight around the world, assisted by a Sperry autopilot that he had nicknamed "Mechanical Mike," the press heralded a new era in aviation. "The days when human skill alone and an almost bird-like sense of direction enabled a flier to hold his course for long hours through a starless night or a fog are over," reported the *New York Times*. "Commercial flying in the future will be automatic."[7]

The introduction of the gyroscopic autopilot set the stage for a momentous expansion of aviation's role in warfare and transport. By taking over much of the manual labor required to keep a plane stable and on course, the device relieved pilots of their constant, exhausting struggle with sticks and pedals, cables and pulleys. That not only alleviated the fatigue aviators endured on long flights; it also freed their hands, their eyes, and, most important, their minds for other, more subtle tasks. They could consult more instruments, make more calculations, solve more problems, and in general think more analytically and creatively about their work. They could fly higher and farther, and with less risk of crashing. They could go out in weather that once would have kept them grounded. And they could undertake intricate maneuvers that would have seemed rash or just plain impossible before. Whether ferrying passengers or dropping bombs, pilots became considerably more versatile and valuable once they had autopilots to help them fly. Their planes changed too: they got bigger, faster, and a whole lot more complicated.

Automatic steering and stabilization tools progressed rapidly during the 1930s, as physicists learned more about aerodynamics and engineers incorporated air-pressure gauges, pneumatic controls, shock absorbers, and other refinements into autopilot mechanisms. The biggest breakthrough came in 1940, when the Sperry Corporation introduced its first electronic model, the A-5. Using vacuum tubes to amplify signals from the gyroscopes, the A-5 was able to make speedier, more precise adjustments and corrections. It could also sense and account for changes in a plane's velocity and acceleration. Used in conjunction with the latest bombsight technology, the electronic autopilot proved a particular boon to the Allied air campaign in World War II.

Shortly after the war, on a September evening in 1947, the U.S. Army Air Forces conducted an experimental flight that made clear how far autopilots had come. Captain Thomas J. Wells, a military test pilot, taxied a C-54 Skymaster transport plane with a seven-man crew onto a remote runway in Newfoundland. He then let go of the yoke, pushed a button to activate the autopilot, and, as one of his colleagues in the cockpit later recalled, "sat back and put his hands in his lap."[8] The plane took off by itself, automatically adjusting its flaps and throttles and, once airborne, retracting its landing gear. It then flew itself across the Atlantic, following a series of "sequences" that had earlier been programmed into what the crew called its "mechanical brain." Each sequence was keyed to a particular altitude or mileage reading. The men on the plane hadn't been told of the flight's route or destination; the plane maintained its own course by monitoring signals from radio beacons on the ground and on boats at sea. At dawn the following day, the C-54 reached the English coast. Still under the control of the autopilot, it began its descent, lowered its landing gear, lined itself up with an airstrip at a Royal Air Force base in Oxfordshire, and executed a perfect landing. Captain Wells then lifted his hands from his lap and parked the plane.

A few weeks after the Skymaster's landmark trip, a writer with the British aviation magazine *Flight* contemplated the implications. It seemed inevitable, he wrote, that the new generation of autopilots would "dispose of the necessity for carrying navigators, radio operators, and flight engineers" on planes. The machines would render those jobs redundant. Pilots, he allowed, did not seem quite so dispensable. They would, at least for the foreseeable future, continue to be a necessary presence in cockpits, if only "to watch the various clocks and indicators to see that everything is going satisfactorily."[9]

■   ■   ■   ■

IN 1988, forty years after the C-54's Atlantic crossing, the European aerospace consortium Airbus Industrie introduced its A320 passenger jet. The 150-seat plane was a smaller version of the company's original A300 model, but unlike its conventional and rather drab predecessor, the A320 was a marvel. The first commercial aircraft that could truly be called computerized, it was a harbinger of everything to come in aircraft design. The flight deck would have been unrecognizable to Wiley Post or Lawrence Sperry. It dispensed with the battery of analogue dials and gauges that had long been the visual signature of airplane cockpits. In their place were six glowing glass screens, of the cathode-ray-tube variety, arranged neatly beneath the windscreen. The displays presented the pilots with the latest data and readings from the plane's network of onboard computers.

The A320's monitor-wrapped flight deck—its "glass cockpit," as pilots called it—was not its most distinctive feature. Engineers at NASA's Langley Research Center had pioneered, more than ten years earlier, the use of CRT screens for transmitting flight information, and jet makers had begun installing the screens in passenger planes in the late 1970s.[10] What really set the A320 apart—and made it, in the words of the American writer and pilot William Langewiesche, "the most audacious civil airplane since the Wright brothers' Flyer"[11]—was its digital fly-by-wire system. Before the A320 arrived, commercial planes still operated mechanically. Their fuse-lages and wing cavities were rigged with cables, pulleys, and gears, along with a miniature waterworks of hydraulic pipes, pumps, and valves. The controls manipulated by a pilot—the yoke, the throttle levers, the rudder pedals—were linked, by means of the mech-anical systems, directly to the moving parts that governed the plane's orientation, direction, and speed. When the pilot acted, the plane reacted.

To stop a bicycle, you squeeze a lever, which pulls a brake cable, which contracts the arms of a caliper, which presses pads against the tire's rim. You are, in essence, sending a command—a signal to stop—with your hand, and the brake mechanism carries the man-ual force of that command all the way to the wheel. Your hand then receives confirmation that your command has been received: you feel, back through the brake lever, the resistance of the caliper, the pressure of the pads against the rim, the skidding of the wheel on the road. That, on a small scale, is what it was like when pilots flew mechanically controlled planes. They became part of the machine, their bodies sensing its workings and feeling its responses, and the machine became a conduit for their will. Such a deep entanglement between human and mechanism was an elemental source of fly-ing's thrill. It's what the famous poet-pilot Antoine de Saint-Exupéry must have had in mind when, in recalling his days flying mail planes in the 1920s, he wrote of how "the machine which at first blush seems a means of isolating man from the great problems of nature, actually plunges him more deeply into them."[12]

The A320's fly-by-wire system severed the tactile link between pilot and plane. It inserted a digital computer between human command and machine response. When a pilot moved a stick, turned a knob, or pushed a button in the Airbus cockpit, his directive was translated, via a transducer, into an electrical signal that zipped down a wire to a computer, and the computer, following the step-by-step algorithms of its software programs, calculated the various mechanical adjustments required to accomplish the pilot's wish. The computer then sent its own instructions to the digital processors that governed the workings of the plane's moving parts. Along with the replacement of mechanical movements by digital signals came a redesign of cockpit controls. The bulky, two-handed yoke that had pulled cables and compressed hydraulic fluids was replaced in the A320 by a small "sidestick" mounted beside the pilot's seat and gripped by one hand. Along the front console, knobs with small, numerical LED displays allowed the pilot to dial in settings for airspeed, altitude, and heading as inputs to the jet's computers.

After the introduction of the A320, the story of airplanes and the story of computers became one. Every advance in hardware and software, in electronic sensors and controls, in display technologies reverberated through the design of commercial aircraft as manufacturers and airlines pushed the limits of automation. In today's jetliners, the autopilots that keep planes stable and on course are just one of many computerized systems. Autothrottles control engine power. Flight management systems gather positioning data from GPS receivers and other sensors and use the information to set or refine a flight path. Collision avoidance systems scan the skies for nearby aircraft. Electronic flight bags store digital copies of the charts and other paperwork that pilots used to carry onboard. Still other computers extend and retract the landing gear, apply the brakes, adjust the cabin pressure, and perform various other functions that had once been in the hands of the crew. To program the computers and monitor their outputs, pilots now use large, colorful flat screens that graphically display data generated by electronic flight instrument systems, along with an assortment of keyboards, keypads, scroll wheels, and other input devices. Computer automation has become "all pervasive" on today's aircraft, says Don Harris, an aeronautics professor and ergonomics expert. The flight deck "can be thought of as one huge flying computer interface."[13]

And what of the modern flyboys and flygirls who, nestled in their high-tech glass cockpits, speed through the air alongside the ghosts of Sperry and Post and Saint-Exupéry? Needless to say, the job of the commercial pilot has lost its aura of romance and adventure. The storied stick-and-rudder man, who flew by a sense of feel, now belongs more to legend than to life. On a typical passenger flight these days, the pilot holds the controls for a grand total of three minutes—a minute or two when taking off and another minute or two when landing. What the pilot spends a whole lot of time doing is checking screens and punching in data. "We've gone from a world where automation was a tool to help the pilot control his workload," observes Bill Voss, president of the Flight Safety Foundation, "to a point where the automation is really the primary flight control system in the aircraft."[14] Writes aviation researcher and FAA advisor Hemant Bhana, "As automation has gained in sophistication, the role of the pilot has shifted toward becoming a monitor or supervisor of the automation."[15] The commercial pilot has become a computer operator. And that, many aviation and automation experts have come to believe, is a problem.

■   ■   ■   ■

LAWRENCE SPERRY died in 1923 when his plane crashed into the English Channel. Wiley Post died in 1935 when his plane went down in Alaska. Antoine de Saint-Exupéry died in 1944 when his plane disappeared over the Mediterranean. Premature death was a routine occupational hazard for pilots during aviation's early years; romance and adventure carried a high price. Passengers died with alarming frequency too. As the airline industry took shape in the 1920s, the publisher of a U.S. aviation journal called on the government to improve flight safety, noting that "a great many fatal accidents are daily occurring to people carried in airplanes by inexperienced pilots."[16]

Air travel's lethal days are, mercifully, behind us. Flying is safe now, and pretty much everyone involved in the aviation business believes that advances in automation are one of the reasons why. Together with improvements in aircraft design, airline safety routines, crew training, and air traffic control, the mechanization and computerization of flight have contributed to the sharp and steady decline in accidents and deaths over the decades. In the United States and other Western countries, fatal airliner crashes have become exceedingly rare. Of the more than seven billion people who boarded U.S. commercial flights in the ten years from 2002 through 2011, only 153 ended up dying in a wreck, a rate of two deaths for every hundred million passengers. In the ten years from 1962 through 1971, by contrast, 1.3 billion people took flights, and 1,696 of them died, for a rate of 133 deaths per hundred million.[17]

But this sunny story carries a dark footnote. The overall decline in the number of plane crashes masks the recent arrival of "a spectacularly new type of accident," says Raja Parasuraman, a psychology professor at George Mason University and one of the world's leading authorities on automation.[18] When onboard computer systems fail to work as intended or other unexpected problems arise during a flight, pilots are forced to take manual control of the plane. Thrust abruptly into a now rare role, they too often make mistakes. The consequences, as the Continental Connection and Air France disasters show, can be catastrophic. Over the last thirty years, dozens of psychologists, engineers, and ergonomics, or "human factors," researchers have studied what's gained and lost when pilots share the work of flying with software. They've learned that a heavy reliance on computer automation can erode pilots' expertise, dull their reflexes, and diminish their attentiveness, leading to what Jan

Noyes, a human-factors expert at Britain's University of Bristol, calls "a deskilling of the crew."[19]

Concerns about the unintended side effects of flight automation aren't new. They date back at least to the early days of glass cockpits and fly-by-wire controls. A 1989 report from NASA's Ames Research Center noted that as computers had begun to multiply on airplanes during the preceding decade, industry and governmental researchers "developed a growing discomfort that the cockpit may be becoming too automated, and that the steady replacement of human functioning by devices could be a mixed blessing." Despite a general enthusiasm for computerized flight, many in the airline industry worried that "pilots were becoming over-dependent on automation, that manual flying skills may be deteriorating, and that situational awareness might be suffering."[20]

Studies conducted since then have linked many accidents and near misses to breakdowns of automated systems or to "automation-induced errors" on the part of flight crews.[21] In 2010, the FAA released preliminary results of a major study of airline flights over the preceding ten years which showed that pilot errors had been involved in nearly two-thirds of all crashes. The research further indicated, according to FAA scientist Kathy Abbott, that automation has made such errors more likely. Pilots can be distracted by their interactions with onboard computers, Abbott said, and they can "abdicate too much responsibility to the automated systems."[22] An extensive 2013 government report on cockpit automation, compiled by an expert panel and drawing on the same FAA data, implicated automation-related problems, such as degraded situational awareness and weakened hand-flying skills, in more than half of recent accidents.[23]

The anecdotal evidence collected through accident reports and surveys gained empirical backing from a rigorous study conducted by Matthew Ebbatson, a young human-factors researcher at Cranfield University, a top U.K. engineering school.[24] Frustrated by the lack

of hard, objective data on what he termed "the loss of manual flying skills in pilots of highly automated airliners," Ebbatson set out
to fill the gap. He recruited sixty-six veteran pilots from a British
airline and had each of them get into a flight simulator and perform a challenging maneuver—bringing a Boeing 737 with a blown
engine in for a landing during bad weather. The simulator disabled
the plane's automated systems, forcing the pilot to fly by hand. Some
of the pilots did exceptionally well in the test, Ebbatson reported,
but many performed poorly, barely exceeding "the limits of accept-
ability." Ebbatson then compared detailed measures of each pilot's
performance in the simulator—the pressure exerted on the yoke,
the stability of airspeed, the degree of variation in course—with the
pilot's historical flight record. He found a direct correlation between
a pilot's aptitude at the controls and the amount of time the pilot
had spent flying without the aid of automation. The correlation was
particularly strong with the amount of manual flying done during
the preceding two months. The analysis indicated that "manual flying skills decay quite rapidly towards the fringes of 'tolerable' performance without relatively frequent practice." Particularly "vulnerable
to decay," Ebbatson noted, was a pilot's ability to maintain "airspeed
control"—a skill crucial to recognizing, avoiding, and recovering
from stalls and other dangerous situations.

It's no mystery why automation degrades pilot performance. Like
many challenging jobs, flying a plane involves a combination of psychomotor skills and cognitive skills—thoughtful action and active
thinking. A pilot needs to manipulate tools and instruments with
precision while swiftly and accurately making calculations, forecasts,
and assessments in his head. And while he goes through these intricate mental and physical maneuvers, he needs to remain vigilant,
alert to what's going on around him and able to distinguish important
signals from unimportant ones. He can't allow himself either to lose

focus or to fall victim to tunnel vision. Mastery of such a multifac-
eted set of skills comes only with rigorous practice. A beginning pilot
tends to be clumsy at the controls, pushing and pulling the yoke with
more force than necessary. He often has to pause to remember what
he should do next, to walk himself methodically through the steps
of a process. He has trouble shifting seamlessly between manual
and cognitive tasks. When a stressful situation arises, he can easily
become overwhelmed or distracted and end up overlooking a critical
change in circumstances.

In time, after much rehearsal, the novice gains confidence. He
becomes less halting in his work and more precise in his actions.
There's little wasted effort. As his experience continues to deepen,
his brain develops so-called mental models—dedicated assem-
blies of neurons—that allow him to recognize patterns in his surroundings. The models enable him to interpret and react to stimuli
intuitively, without getting bogged down in conscious analysis.
Eventually, thought and action become seamless. Flying becomes
second nature. Years before researchers began to plumb the workings of pilots' brains, Wiley Post described the experience of expert
flight in plain, precise terms. He flew, he said in 1935, "without
mental effort, letting my actions be wholly controlled by my subconscious mind."[25] He wasn't born with that ability. He developed it
through hard work.

When computers enter the picture, the nature and the rigor of the
work change, as does the learning the work engenders. As software
assumes moment-by-moment control of the craft, the pilot is, as we've
seen, relieved of much manual labor. This reallocation of responsibility can provide an important benefit. It can reduce the pilot's workload and allow him to concentrate on the cognitive aspects of flight.
But there's a cost. Psychomotor skills get rusty, which can hamper the
pilot on those rare but critical occasions when he's required to take

back the controls. There's growing evidence that recent expansions in the scope of automation also put cognitive skills at risk. When more advanced computers begin to take over planning and analysis functions, such as setting and adjusting a flight plan, the pilot becomes less engaged not only physically but mentally. Because the precision and speed of pattern recognition appear to depend on regular practice, the pilot's mind may become less agile in interpreting and reacting to fast-changing situations. He may suffer what Ebbatson calls "skill fade" in his mental as well as his motor abilities.

Pilots are not blind to automation's toll. They've always been wary about ceding responsibility to machinery. Airmen in World War I, justifiably proud of their skill in maneuvering their planes during dogfights, wanted nothing to do with the fancy Sperry autopilots.[26] In 1959, the original Mercury astronauts rebelled against NASA's plan to remove manual flight controls from spacecraft.[27] But aviators' concerns are more acute now. Even as they praise the enormous gains in flight technology, and acknowledge the safety and efficiency benefits, they worry about the erosion of their talents. As part of his research, Ebbatson surveyed commercial pilots, asking them whether "they felt their manual flying ability had been influenced by the experience of operating a highly automated aircraft." More than three-fourths reported that "their skills had deteriorated"; just a few felt their skills had improved.[28] A 2012 pilot survey conducted by the European Aviation Safety Agency found similarly widespread concerns, with 95 percent of pilots saying that automation tended to erode "basic manual and cognitive flying skills."[29] Rory Kay, a longtime United Airlines captain who until recently served as the top safety official with the Air Line Pilots Association, fears the aviation industry is suffering from "automation addiction." In a 2011 interview with the Associated Press, he put the problem in stark terms: "We're forgetting how to fly."[30]

■   ■   ■   ■

CYNICS ARE quick to attribute such fears to self-interest. The real reason for the grumbling about automation, they contend, is that pilots are anxious about the loss of their jobs or the squeezing of their paychecks. And the cynics are right, to a degree. As the writer for *Flight* magazine predicted back in 1947, automation technology has whittled down the size of flight crews. Sixty years ago, an airliner's flight deck often had seats for five skilled and well-paid professionals: a navigator, a radio operator, a flight engineer, and a pair of pilots. The radioman lost his chair during the 1950s, as communication systems became more reliable and easier to use. The navigator was pushed off the deck in the 1960s, when inertial navigation systems took over his duties. The flight engineer, whose job involved monitoring a plane's instrument array and relaying important information to the pilots, kept his seat until the advent of the glass cockpit at the end of the 1970s. Seeking to cut costs following the deregulation of air travel in 1978, American airlines made a push to get rid of the engineer and fly with just a captain and copilot. A bitter battle with pilots' unions ensued, as the unions mobilized to save the engineer's job. The fight didn't end until 1981, when a U.S. presidential commission declared that engineers were no longer necessary for the safe operation of passenger flights. Since then, the two-person flight crew has become the norm—at least for the time being. Some experts, pointing to the success of military drones, have begun suggesting that two pilots may in the end be two too many.[31] "A pilotless airliner is going to come," James Albaugh, a top Boeing executive, told an aviation conference in 2011; "it's just a question of when."[32]

The spread of automation has also been accompanied by a steady decline in the compensation of commercial pilots. While veteran jetliner captains can still pull down salaries close to $200,000, novice pilots today are paid as little as $20,000 a year, sometimes even less. The average starting salary for experienced pilots at major airlines is around $36,000, which, as a *Wall Street Journal* reporter notes, is "darn low for mid-career professionals."[33] Despite the modest pay, there's still a popular sense that pilots are overcompensated. An article at the website Salary.com called commercial jet pilots the "most overpaid" professionals in today's economy, arguing that "many of their tasks are automated" and suggesting their work has become "a bit boring."[34]

But pilots' self-interest, when it comes to matters of automation, goes deeper than employment security and pay, or even their own safety. Every technological advance alters the work they do and the role they play, and that in turn changes how they view themselves and how others see them. Their social status and even their sense of self are in play. So when pilots talk about automation, they're speaking not just technically but autobiographically. Am I the master of the machine, or its servant? Am I an actor in the world, or an observer? Am I an agent, or an object? "At heart," MIT technology historian David Mindell writes in his book *Digital Apollo*, "debates about control and automation in aircraft are debates about the relative importance of human and machine." In aviation, as in any field where people work with tools, "technical change and social change are intertwined."[35]

Pilots have always defined themselves by their relationship to their craft. Wilbur Wright, in a 1900 letter to Octave Chanute, another aviation pioneer, said of the pilot's role, "What is chiefly needed is skill rather than machinery."[36] He was not just voicing a platitude. He was referring to what, at the very dawn of human flight, had already become a crucial tension between the capability of the plane and the capability of the pilot. As the first planes were being built, designers debated how inherently stable an aircraft should be—how strong of a tendency it should have to fly straight and level in all conditions. It might seem that more stability would always be better in a flying machine, but that's not so. There's a trade-off between stability and maneuverability. The greater a plane's stability, the harder it becomes for the pilot to exert control over it. As Mindell explains, "The more stable an aircraft is, the more effort will be required to move it off its point of equilibrium. Hence it will be less controllable. The opposite is also true—the more controllable, or maneuverable, an aircraft, the less stable it will be."[37] The author of a 1910 book on aeronautics reported that the question of equilibrium had become "a controversy dividing aviators into two schools." On one side were those who argued that equilibrium should "be made automatic to a very large degree"—that it should be built into the plane. On the other side were those who held that equilibrium should be "a matter for the skill of the aviator."[38]

Wilbur and Orville Wright were in the latter camp. They believed that a plane should be fundamentally unstable, like a bicycle or even, as Wilbur once suggested, "a fractious horse."[39] That way, the pilot would have as much autonomy and freedom as possible. The brothers incorporated their philosophy into the planes they built, which gave precedence to maneuverability over stability. What the Wrights invented at the start of the twentieth century was, Mindell argues, "not simply an airplane that could fly, but also the *very idea* of an airplane as a dynamic machine under the control of a human pilot."[40] Before the engineering decision came an ethical choice: to make the apparatus subservient to the person operating it, an instrument of human talent and volition.

The Wright brothers would lose the equilibrium debate. As planes came to carry passengers and other valuable cargo over long distances, the freedom and virtuosity of the pilot became secondary concerns. Of primary importance were safety and efficiency, and to increase those, it quickly became clear, the pilot's scope of action had to be constrained and the machine itself invested with more authority. The shift in control was gradual, but every time technology assumed a little more power, pilots felt a little more of themselves slip away. In a quixotic 1957 article opposing attempts to further automate flight, a top fighter-jet test pilot named J. O. Roberts fretted about how autopilots were turning the man in the cockpit into little more than "excess baggage except for monitoring duties." The pilot, Roberts wrote, has to wonder "whether he is paying his way or not."[41]

But all the gyroscopic, electromechanical, instrumental, and hydraulic innovations only hinted at what digitization would bring. The computer not only changed the character of flight; it changed the character of automation. It circumscribed the pilot's role to the point where the very idea of "manual control" began to seem anachronistic. If the essence of a pilot's job consists in sending digital inputs to computers and monitoring computers' digital outputs—while the computers govern the plane's moving parts and choose its course—where exactly is the manual control? Even when pilots in computerized planes are pulling yokes or pushing sticks, what they're often really involved in is a simulation of manual flight. Every action is mediated, filtered through microprocessors. That's not to say that there aren't still important skills involved. There are. But the skills have changed, and they're now applied at a distance, from behind a scrim of software. In many of today's commercial jets, the flight software can even override the pilots' inputs during extreme maneuvers. The computer gets the final say. "He didn't just fly an airplane," a fellow pilot once said of Wiley Post; "he put it on."[42] Today's pilots don't wear their planes. They wear their planes' computers—or perhaps the computers wear the pilots.

The transformation that aviation has gone through over the last few decades—the shift from mechanical to digital systems, the proliferation of software and screens, the automation of mental as well as manual work, the blurring of what it means to be a pilot—offers a roadmap for the much broader transformation that society is going through now. The glass cockpit, Don Harris has pointed out, can be thought of as a prototype of a world where "there is computer functionality everywhere."[43] The experience of pilots also reveals the subtle but often strong connection between the way automated systems are designed and the way the minds and bodies of the people using the systems work. The mounting evidence of an erosion of skills, a dulling of perceptions, and a slowing of reactions should give us all pause. As we begin to live our lives inside glass cockpits, we seem fated to discover what pilots already know: a glass cockpit can also be a glass cage.

CHAPTER FOUR

THE DEGENERATION EFFECT

A HUNDRED YEARS AGO, IN HIS BOOK *AN INTRODUCTION to Mathematics*, the British philosopher Alfred North Whitehead wrote, "Civilization advances by extending the number of important operations which we can perform without thinking about them." Whitehead wasn't writing about machinery. He was writing about the use of mathematical symbols to represent ideas or logical processes— an early example of how intellectual work can be encapsulated in code. But he intended his observation to be taken generally. The common notion that "we should cultivate the habit of thinking of what we are doing," he wrote, is "profoundly erroneous." The more we can relieve our minds of routine chores, offloading the tasks to technological aids, the more mental power we'll be able to store up for the deepest, most creative kinds of reasoning and conjecture. "Operations of thought are like cavalry charges in battle—they are strictly limited in number, they require fresh horses, and must only be made at decisive moments."[1]

It's hard to imagine a more succinct or confident expression of faith in automation as a cornerstone of progress. Implicit in Whitehead's words is a belief in a hierarchy of human action. Every time we offload a job to a tool or a machine, or to a symbol or a software algorithm, we free ourselves to climb to a higher pursuit, one requiring greater dexterity, richer intelligence, or a broader perspective. We may lose something with each upward step, but what we gain is, in the end, far greater. Taken to an extreme, Whitehead's sense of automation as liberation turns into the techno-utopianism of Wilde and Keynes, or Marx at his sunniest—the dream that machines will free us from our earthly labors and deliver us back to an Eden of leisurely delights. But Whitehead didn't have his head in the clouds. He was making a pragmatic point about how to spend our time and exert our effort. In a publication from the 1970s, the U.S. Department of Labor summed up the job of secretaries by saying that they "relieve their employers of routine duties so they can work on more important matters."[2] Software and other automation technologies, in the Whitehead view, play an analogous role.

History provides plenty of evidence to support Whitehead. People have been handing off chores, both physical and mental, to tools since the invention of the lever, the wheel, and the counting bead. The transfer of work has allowed us to tackle thornier challenges and rise to greater achievements. That's been true on the farm, in the factory, in the laboratory, in the home. But we shouldn't take Whitehead's observation for a universal truth. He was writing when automation was limited to distinct, well-defined, and repetitive tasks—weaving fabric with a steam loom, harvesting grain with a combine, multiplying numbers with a slide rule. Automation is different now. Computers, as we've seen, can be programmed to perform or support complex activities in which a succession of tightly coordinated tasks is carried out through an evaluation of many variables. In automated systems today, the computer often takes on intellectual work—observing and sensing, analyzing and judging, even making decisions—that until recently was considered the preserve of humans. The person operating the computer is left to play the role of a high-tech clerk, entering data, monitoring outputs, and watching for failures. Rather than opening new frontiers of thought and action to its human collaborators, software narrows our focus. We trade subtle, specialized talents for more routine, less distinctive ones.

Most of us assume, as Whitehead did, that automation is benign, that it raises us to higher callings but doesn't otherwise alter the way we behave or think. That's a fallacy. It's an expression of what scholars of automation have come to call the "substitution myth." A labor-saving device doesn't just provide a substitute for some isolated component of a job. It alters the character of the entire task, including the roles, attitudes, and skills of the people who take part in it. As Raja Parasuraman explained in a 2000 journal article, "Automation does not simply supplant human activity but rather changes it, often in ways unintended and unanticipated by the designers."[3] Automation remakes both work and worker.

■   ■   ■   ■

WHEN PEOPLE tackle a task with the aid of computers, they often fall victim to a pair of cognitive ailments, *automation complacency* and *automation bias*. Both reveal the traps that lie in store when we take the Whitehead route of performing important operations without thinking about them.

Automation complacency takes hold when a computer lulls us into a false sense of security. We become so confident that the machine will work flawlessly, handling any challenge that may arise, that we allow our attention to drift. We disengage from our work, or at least from the part of it that the software is handling, and as a result may miss signals that something is amiss. Most of us have experienced complacency when at a computer. In using email or word-processing software, we become less vigilant proofreaders when the spell checker is on.[4] That's a simple example, which at worst can lead to a moment of embarrassment. But as the sometimes tragic experience of avia-tors shows, automation complacency can have deadly consequences. In the worst cases, people become so trusting of the technology that their awareness of what's going on around them fades completely. They tune out. If a problem suddenly crops up, they may act bewil-dered and waste precious moments trying to reorient themselves.

Automation complacency has been documented in many high-risk situations, from battlefields to industrial control rooms to the bridges of ships and submarines. One classic case involved a 1,500-passenger ocean liner named the *Royal Majesty*, which in the spring of 1995 was sailing from Bermuda to Boston on the last leg of a week-long cruise. The ship was outfitted with a state-of-the-art automated navigation system that used GPS signals to keep it on course. An hour into the voyage, the cable for the GPS antenna came loose and the navigation system lost its bearings. It continued to give readings, but they were no longer accurate. For more than thirty hours, as the ship slowly drifted off its appointed route, the captain and crew remained oblivi-ous to the problem, despite clear signs that the system had failed. At one point, a mate on watch was unable to spot an important loca-tional buoy that the ship was due to pass. He failed to report the fact. His trust in the navigation system was so complete that he assumed the buoy was there and he simply didn't see it. Nearly twenty miles off course, the ship finally ran aground on a sandbar near Nantucket Island. No one was hurt, fortunately, though the cruise company suffered millions in damages. Government safety investigators con-cluded that automation complacency caused the mishap. The ship's officers were "overly reliant" on the automated system, to the point that they ignored other "navigation aids [and] lookout information" that would have told them they were dangerously off course. Automa-tion, the investigators reported, had "the effect of leaving the mariner out of meaningful control or active participation in the operation of the ship."[5]

Complacency can plague people who work in offices as well as those who ply airways and seaways. In an investigation of how design software has influenced the building trades, MIT sociologist Sherry Turkle documented a change in architects' attention to detail. When plans were hand-drawn, architects would painstakingly double-check all the dimensions before handing blueprints over to construction crews. The architects knew that they were fallible, that they could make the occasional goof, and so they followed an old carpentry dictum: measure twice, cut once. With software-generated plans, they're less careful about verifying measurements. The apparent precision of computer renderings and printouts leads them to assume that the figures are accurate. "It seems presumptuous to check," one architect told Turkle; "I mean, how could I do a better job than the computer? It can do things down to hundredths of an inch." Such complacency, which can be shared by engineers and builders, has led to costly mistakes in planning and construction. Computers don't make goofs, we tell ourselves, even though we know that their outputs are only as good as our inputs. "The fancier the computer system," one of Turkle's students observed, "the more you start to assume that it is correcting your errors, the more you start to believe that what comes out of the machine is just how it should be. It is just a visceral thing."[6]

Automation bias is closely related to automation complacency. It creeps in when people give undue weight to the information coming through their monitors. Even when the information is wrong or misleading, they believe it. Their trust in the software becomes so strong that they ignore or discount other sources of information, including their own senses. If you've ever found yourself lost or going around in circles after slavishly following flawed or outdated directions from a GPS device or other digital mapping tool, you've felt the effects of automation bias. Even people who drive for a living can display a startling lack of common sense when relying on satellite navigation. Ignoring road signs and other environmental cues, they'll proceed down hazardous routes and sometimes end up crashing into low overpasses or getting stuck in the narrow streets of small towns. In Seattle in 2008, the driver of a twelve-foot-high bus carrying a high-school sports team ran into a concrete bridge with a nine-foot clearance. The top of the bus was sheared off, and twenty-one injured students had to be taken to the hospital. The driver told police that he had been following GPS instructions and "did not see" signs and flashing lights warning of the low bridge ahead.[7]

Automation bias is a particular risk for people who use decision-support software to guide them through analyses or diagnoses. Since the late 1990s, radiologists have been using computer-aided detection systems that highlight suspicious areas on mammograms and other x-rays. A digital version of an image is scanned into a computer, and pattern-matching software reviews it and adds arrows or other "prompts" to suggest areas for the doctor to inspect more closely. In some cases, the highlights aid in the discovery of disease, helping radiologists identify potential cancers they might otherwise have missed. But studies reveal that the highlights can also have the opposite effect. Biased by the software's suggestions, doctors can end up giving cursory attention to the areas of an image that haven't been highlighted, sometimes overlooking an early-stage tumor or other abnormality. The prompts can also increase the likelihood of false-positives, when a radiologist calls a patient back for an unnecessary biopsy.

A recent review of mammography data, conducted by a team of researchers at City University London, indicates that automation bias has had a greater effect on radiologists and other image readers than was previously thought. The researchers found that while computer-aided detection tends to improve the reliability of "less discriminating readers" in assessing "comparatively easy cases," it can actually degrade the performance of expert readers in evaluating tricky cases. When relying on the software, the experts are more likely to overlook certain cancers.[8] The subtle biases inspired by computerized decision aids may, moreover, be "an inherent part of the human cognitive apparatus for reacting to cues and alarms."[9] By directing the focus of our eyes, the aids distort our vision.

Both complacency and bias seem to stem from limitations in our ability to pay attention. Our tendency toward complacency reveals how easily our concentration and awareness can fade when we're not routinely called on to interact with our surroundings. Our propensity to be biased in evaluating and weighing information shows that our mind's focus is selective and can easily be skewed by misplaced trust or even the appearance of seemingly helpful prompts. Both complacency and bias tend to become more severe as the quality and reliability of an automated system improve.[10] Experiments show that when a system produces errors fairly frequently, we stay on high alert. We maintain awareness of our surroundings and carefully monitor information from a variety of sources. But when a system is more reliable, breaking down or making mistakes only occasionally, we get lazy. We start to assume the system is infallible.

Because automated systems usually work fine even when we lose awareness or objectivity, we are rarely penalized for our complacency or our bias. That ends up compounding the problems, as Parasuraman pointed out in a 2010 paper written with his German colleague Dietrich Manzey. "Given the usually high reliability of automated systems, even highly complacent and biased behavior of operators rarely leads to obvious performance consequences," the scholars wrote. The lack of negative feedback can in time induce "a cognitive process that resembles what has been referred to as 'learned carelessness.'"[11] Think about driving a car when you're sleepy. If you begin to nod off and drift out of your lane, you'll usually go onto a rough shoulder, hit a rumble strip, or earn a honk from another motorist—signals that jolt you back awake. If you're in a car that automatically keeps you within a lane by monitoring the lane markers and adjusting the steering, you won't receive such warnings. You'll drift into a deeper slumber. Then if something unexpected happens—an animal runs into the road, say, or a car stops short in front of you—you'll be much more likely to have an accident. By isolating us from negative feedback, automation makes it harder for us to stay alert and engaged. We tune out even more.

■   ■   ■   ■

OUR SUSCEPTIBILITY to complacency and bias explains how a reliance on automation can lead to errors of both commission and omission. We accept and act on information that turns out to be incorrect or incomplete, or we fail to see things that we should have seen. But the way that a reliance on computers weakens awareness and attentiveness also points to a more insidious problem. Automation tends to turn us from actors into observers. Instead of manipulating the yoke, we watch the screen. That shift may make our lives easier, but it can also inhibit our ability to learn and to develop expertise. Whether automation enhances or degrades our performance in a given task, over the long run it may diminish our existing skills or prevent us from acquiring new ones.

Since the late 1970s, cognitive psychologists have been documenting a phenomenon called the generation effect. It was first observed in studies of vocabulary, which revealed that people remember words much better when they actively call them to mind—when they *generate* them—than when they read them from a page. In one early and famous experiment, conducted by University of Toronto psychologist Norman Slamecka, people used flash cards to memorize pairs of antonyms, like *hot* and *cold*. Some of the test subjects were given cards that had both words printed in full, like this:

HOT : COLD

Others used cards that showed only the first letter of the second word, like this:

HOT : C

The people who used the cards with the missing letters performed much better in a subsequent test measuring how well they remembered the word pairs. Simply forcing their minds to fill in a blank, to act rather than observe, led to stronger retention of information.[12]

The generation effect, it has since become clear, influences memory and learning in many different circumstances. Experiments have revealed evidence of the effect in tasks that involve not only remembering letters and words but also remembering numbers, pictures, and sounds, completing math problems, answering trivia questions, and reading for comprehension. Recent studies have also demonstrated the benefits of the generation effect for higher forms of teaching and learning. A 2011 paper in *Science* showed that students who read a complex science assignment during a study period and then spent a second period recalling as much of it as possible, unaided, learned the material more fully than students who read the assignment repeatedly over the course of four study periods.[13] The mental act of generation improves people's ability to carry out activities that, as education researcher Britte Haugan Cheng has written, "require conceptual reasoning and requisite deeper cognitive processing." Indeed, Cheng says, the generation effect appears to strengthen as the material generated by the mind becomes more complex.[14]

Psychologists and neuroscientists are still trying to figure out what goes on in our minds to give rise to the generation effect. But it's clear that deep cognitive and memory processes are involved. When we work hard at something, when we make it the focus of attention and effort, our mind rewards us with greater understanding. We remember more and we learn more. In time, we gain know-how, a particular talent for acting fluidly, expertly, and purposefully in the world. That's hardly a surprise. Most of us know that the only way to get good at something is by actually doing it. It's easy to gather information quickly from a computer screen—or from a book, for that matter. But true knowledge, particularly the kind that lodges deep in memory and manifests itself in skill, is harder to come by. It requires a vigorous, prolonged struggle with a demanding task.

The Australian psychologists Simon Farrell and Stephan Lewandowsky made the connection between automation and the generation effect in a paper published in 2000. In Slamecka's experiment, they pointed out, supplying the second word of an antonym pair, rather than forcing a person to call the word to mind, "can be considered an instance of automation because a human activity—generation of the word 'COLD' by participants—has been obviated by a printed stimulus." By extension, "the reduction in performance that is observed when generation is replaced by reading can be considered a manifestation of complacency."[15] That helps illuminate the cognitive cost of automation. When we carry out a task or a job on our own, we seem to use different mental processes than when we rely on the aid of a computer. When software reduces our engagement with our work, and in particular when it pushes us into a more passive role as observer or monitor, we circumvent the deep cognitive processing that underpins the generation effect. As a result, we hamper our ability to gain the kind of rich, real-world knowledge that leads to know-how. The generation effect requires precisely the kind of struggle that automation seeks to alleviate.

In 2004, Christof van Nimwegen, a cognitive psychologist at Utrecht University in the Netherlands, began a series of simple but ingenious experiments to investigate software's effects on memory formation and the development of expertise.[16] He recruited two groups of people and had them play a computer game based on a classic logic puzzle called Missionaries and Cannibals. To complete the puzzle, a player has to transport across a hypothetical river five missionaries and five cannibals (or, in van Nimwegen's version, five yellow balls and five blue ones), using a boat that can accommodate no more than three passengers at a time. The tricky part is that there can never be more cannibals than missionaries in one place, either in the boat or on the riverbanks. (If outnumbered, the missionaries become the cannibals' dinner, one assumes.) Figuring out the series of boat trips that can best accomplish the task requires rigorous analysis and careful planning.

One of van Nimwegen's groups worked on the puzzle using software that provided step-by-step guidance, offering, for instance, on-screen prompts to highlight which moves were permissible and which weren't. The other group used a rudimentary program that offered no assistance. As you'd expect, the people using the helpful software made faster progress at the outset. They could follow the prompts rather than having to pause before each move to recall the rules and figure out how they applied to the new situation. But as the game advanced, the players using the rudimentary software began to excel. In the end, they were able to work out the puzzle more efficiently, with significantly fewer wrong moves, than their counterparts who were receiving assistance. In his report on the experiment, van Nimwegen concluded that the subjects using the rudimentary program developed a clearer conceptual understanding of the task. They were better able to think ahead and plot a successful strategy. Those relying on guidance from the software, by contrast, often became confused and would "aimlessly click around."

The cognitive penalty imposed by the software aids became even clearer eight months later, when van Nimwegen had the same people work through the puzzle again. Those who had earlier used the rudimentary software finished the game almost twice as quickly as their counterparts. The subjects using the basic program, he wrote, displayed "more focus" during the task and "better imprinting of knowledge" afterward. They enjoyed the benefits of the generation effect. Van Nimwegen and some of his Utrecht colleagues went on to conduct experiments involving more realistic tasks, such as using calendar software to schedule meetings and event-planning software to assign conference speakers to rooms. The results were the same. People who relied on the help of software prompts displayed less strategic thinking, made more superfluous moves, and ended up with a weaker conceptual understanding of the assignment. Those using unhelpful programs planned better, worked smarter, and learned more.[17]

What van Nimwegen observed in his laboratory—that when we automate cognitive tasks like problem solving, we hamper the mind's ability to translate information into knowledge and knowledge into know-how—is also being documented in the real world. In many businesses, managers and other professionals depend on so-called expert systems to sort and analyze information and suggest courses of action. Accountants, for example, use decision-support software in corporate audits. The applications speed the work, but there are signs that as the software becomes more capable, the accountants become less so. One study, conducted by a group of Australian professors, examined the effects of the expert systems used by three international accounting firms. Two of the companies employed advanced software that, based on an accountant's answers to basic questions about a client, recommended a set of relevant business risks to include in the client's audit file. The third firm used simpler software that provided a list of potential risks but required the accountant to review them and manually select the pertinent ones for the file. The researchers gave accountants from each firm a test measuring their knowledge of risks in industries in which they had performed audits. Those from the firm with the less helpful software displayed a significantly stronger understanding of different forms of risk than did those from the other two firms. The decline in learning associated with advanced software affected even veteran auditors—those with more than five years of experience at their current firm.[18]

Other studies of expert systems reveal similar effects. The research indicates that while decision-support software can help novice analysts make better judgments in the short run, it can also make them mentally lazy. By diminishing the intensity of their thinking, the software retards their ability to encode information in memory, which makes them less likely to develop the rich tacit knowledge essential to true expertise.[19] The drawbacks to automated decision aids can be subtle, but they have real consequences, particularly in fields where analytical errors have far-reaching repercussions. Miscalculations of risk, exacerbated by high-speed computerized trading programs, played a major role in the near meltdown of the world's financial system in 2008. As Tufts University management professor Amar Bhidé has suggested, "robotic methods" of decision making led to a widespread "judgment deficit" among bankers and other Wall Street professionals.[20] While it may be impossible to pin down the precise degree to which automation figured in the disaster, or in subsequent fiascos like the 2010 "flash crash" on U.S. exchanges, it seems prudent to take seriously any indication that a widely used technology may be diminishing the knowledge or clouding the judgment of people in sensitive jobs. In a 2013 paper, computer scientists Gordon Baxter and John Cartlidge warned that a reliance on automation is eroding the skills and knowledge of financial professionals even as computer-trading systems make financial markets more risky.[21]

Some software writers worry that their profession's push to ease

the strain of thinking is taking a toll on their own skills. Programmers today often use applications called integrated development
environments, or IDEs, to aid them in composing code. The applications automate many tricky and time-consuming chores. They
typically incorporate auto-complete, error-correction, and debugging
routines, and the more sophisticated of them can evaluate and revise
the structure of a program through a process known as refactoring.
But as the applications take over the work of coding, programmers
lose opportunities to practice their craft and sharpen their talent.
"Modern IDEs are getting 'helpful' enough that at times I feel like an
IDE operator rather than a programmer," writes Vivek Haldar, a veteran software developer with Google. "The behavior all these tools
encourage is not 'think deeply about your code and write it carefully,'
but 'just write a crappy first draft of your code, and then the tools will
tell you not just what's wrong with it, but also how to make it better.'"
His verdict: "Sharp tools, dull minds."[22]

Google acknowledges that it has even seen a dumbing-down
effect among the general public as it has made its search engine
more responsive and solicitous, better able to predict what people
are looking for. Google does more than correct our typos; it suggests search terms as we type, untangles semantic ambiguities in
our requests, and anticipates our needs based on where we are and
how we've behaved in the past. We might assume that as Google gets
better at helping us refine our searching, we would learn from its
example. We would become more sophisticated in formulating keywords and otherwise honing our online explorations. But according
to the company's top search engineer, Amit Singhal, the opposite is
the case. In 2013, a reporter from the *Observer* newspaper in London
interviewed Singhal about the many improvements that have been
made to Google's search engine over the years. "Presumably," the
journalist remarked, "we have got more precise in our search terms
the more we have used Google." Singhal sighed and, "somewhat wea-

rily," corrected the reporter: "'Actually, it works the other way. The more accurate the machine gets, the lazier the questions become.'"[23]

More than our ability to compose sophisticated queries may be compromised by the ease of search engines. A series of experiments reported in *Science* in 2011 indicates that the ready availability of information online weakens our memory for facts. In one of the experiments, test subjects read a few-dozen simple, true statements—"an ostrich's eye is bigger than its brain," for instance—and then typed them into a computer. Half the subjects were told the computer would save what they typed; the other half were told that the statements would be erased. Afterward, the participants were asked to write down all the statements they could recall. People who believed the information had been stored in the computer remembered significantly fewer of the facts than did those who assumed the statements had not been saved. Just knowing that information will be available in a database appears to reduce the likelihood that our brains will make the effort required to form memories. "Since search engines are continually available to us, we may often be in a state of not feeling we need to encode the information internally," the researchers concluded. "When we need it, we will look it up."[24]

For millennia, people have supplemented their biological memory with storage technologies, from scrolls and books to microfiche and magnetic tape. Tools for recording and distributing information underpin civilization. But external storage and biological memory are not the same thing. Knowledge involves more than looking stuff up; it requires the encoding of facts and experiences in personal memory. To truly know something, you have to weave it into your neural circuitry, and then you have to repeatedly retrieve it from memory and put it to fresh use. With search engines and other online resources, we've automated information storage and retrieval to a degree far beyond anything seen before. The brain's seemingly innate tendency to offload, or externalize, the work of remembering makes us more efficient thinkers in some ways. We can quickly call up facts that have slipped our mind. But that same tendency can become pathological when the automation of mental labor makes it too easy to avoid the work of remembering and understanding.

Google and other software companies are, of course, in the business of making our lives easier. That's what we ask them to do, and it's why we're devoted to them. But as their programs become adept at doing our thinking for us, we naturally come to rely more on the software and less on our own smarts. We're less likely to push our minds to do the work of generation. When that happens, we end up learning less and knowing less. We also become less capable. As the University of Texas computer scientist Mihai Nadin has observed, in regard to modern software, "The more the interface replaces human effort, the lower the adaptivity of the user to new situations."[25] In place of the generation effect, computer automation gives us the reverse: a degeneration effect.

■   ■   ■   ■

BEAR WITH me while I draw your attention back to that ill-fated, slicker-yellow Subaru with the manual transmission. As you'll recall, I went from hapless gear-grinder to reasonably accomplished stick-handler with just a few weeks' practice. The arm and leg movements my dad had taught me, cursorily, now seemed instinctive. I was hardly an expert, but shifting was no longer a struggle. I could do it without thinking. It had become, well, automatic.

My experience provides a model for the way humans gain complicated skills. We often start off with some basic instruction, received directly from a teacher or mentor or indirectly from a book or manual or YouTube video, which transfers to our conscious mind explicit knowledge about how a task is performed: do this, then this, then this. That's what my father did when he showed me the location of the gears and explained when to step on the clutch. As I quickly discovered, explicit knowledge goes only so far, particularly when the task has a psychomotor component as well as a cognitive one. To achieve mastery, you need to develop tacit knowledge, and that comes only through real experience—by rehearsing a skill, over and over again. The more you practice, the less you have to think about what you're doing. Responsibility for the work shifts from your conscious mind, which tends to be slow and halting, to your unconscious mind, which is quick and fluid. As that happens, you free your conscious mind to focus on the more subtle aspects of the skill, and when those, too, become automatic, you proceed up to the next level. Keep going, keep pushing yourself, and ultimately, assuming you have some native aptitude for the task, you're rewarded with expertise.

This skill-building process, through which talent comes to be exercised without conscious thought, goes by the ungainly name *automatization*, or the even more ungainly name *proceduralization*. Automatization involves deep and widespread adaptations in the brain. Certain brain cells, or neurons, become fine-tuned for the task at hand, and they work in concert through the electrochemical connections provided by synapses. The New York University cognitive psychologist Gary Marcus offers a more detailed explanation: "At the neural level, proceduralization consists of a wide array of carefully coordinated processes, including changes to both gray matter (neural cell bodies) and white matter (axons and dendrites that connect between neurons). Existing neural connections (synapses) must be made more efficient, new dendritic spines may be formed, and proteins must be synthesized."[26] Through the neural modifications of automatization, the brain develops *automaticity*, a capacity for rapid, unconscious perception, interpretation, and action that allows mind and body to recognize patterns and respond to changing circumstances instantaneously.

All of us experienced automatization and achieved automaticity when we learned to read. Watch a young child in the early stages of reading instruction, and you'll witness a taxing mental struggle. The child has to identify each letter by studying its shape. She has to sound out how a set of letters combine to form a syllable and how a series of syllables combine to form a word. If she's not already familiar with the word, she has to figure out or be told its meaning. And then, word by word, she has to interpret the meaning of a sentence, often resolving the ambiguities inherent to language. It's a slow, painstaking process, and it requires the full attention of the conscious mind. Eventually, though, letters and then words get encoded in the neurons of the visual cortex—the part of the brain that processes sight—and the young reader begins to recognize them without conscious thought. Through a symphony of brain changes, reading becomes effortless. The greater the automaticity the child achieves, the more fluent and accomplished a reader she becomes.[27]

Whether it's Wiley Post in a cockpit, Serena Williams on a tennis court, or Magnus Carlsen at a chessboard, the otherworldly talent of the virtuoso springs from automaticity. What looks like instinct is hard-won skill. Those changes in the brain don't happen through passive observation. They're generated through repeated confrontations with the unexpected. They require what the philosopher of mind Hubert Dreyfus terms "experience in a variety of situations, all seen from the same perspective but requiring different tactical decisions."[28] Without lots of practice, lots of repetition and rehearsal of a skill in different circumstances, you and your brain will never get really good at anything, at least not anything complicated. And without continuing practice, any talent you do achieve will get rusty.

It's popular now to suggest that practice is all you need. Work at a skill for ten thousand hours or so, and you'll be blessed with expertise—you'll become the next great pastry chef or power forward. That, unhappily, is an exaggeration. Genetic traits, both physical and intellectual, do play an important role in the development of talent, particularly at the highest levels of achievement. Nature matters. Even our desire and aptitude for practice has, as Marcus points out, a genetic component: "How we respond to experience, and even what type of experience we seek, are themselves in part functions of the genes we are born with."[29] But if genes establish, at least roughly, the upper bounds of individual talent, it's only through practice that a person will ever reach those limits and fulfill his or her potential. While innate abilities make a big difference, write psychology professors David Hambrick and Elizabeth Meinz, "research has left no doubt that one of the largest sources of individual differences in performance on complex tasks is simply what and how much people know: declarative, procedural, and strategic knowledge acquired through years of training and practice in a domain."[30]

Automaticity, as its name makes clear, can be thought of as a kind of internalized automation. It's the body's way of making difficult but repetitive work routine. Physical movements and procedures get programmed into muscle memory; interpretations and judgments are made through the instant recognition of environmental patterns apprehended by the senses. The conscious mind, scientists discovered long ago, is surprisingly cramped, its capacity for taking in and processing information limited. Without automaticity, our consciousness would be perpetually overloaded. Even very simple acts, such as reading a sentence in a book or cutting a piece of steak with a knife and fork, would strain our cognitive capabilities. Automaticity gives us more headroom. It increases, to put a different spin on Alfred North Whitehead's observation, "the number of important operations which we can perform without thinking about them."

Tools and other technologies, at their best, do something similar, as Whitehead appreciated. The brain's capacity for automaticity has limits of its own. Our unconscious mind can perform a lot of functions quickly and efficiently, but it can't do everything. You might be able to memorize the times table up to twelve or even twenty, but you would probably have trouble memorizing it much beyond that. Even if your brain didn't run out of memory, it would probably run out of patience. With a simple pocket calculator, though, you can automate even very complicated mathematical procedures, ones that would tax your unaided brain, and free up your conscious mind to consider what all that math adds up to. But that only works if you've already mastered basic arithmetic through study and practice. If you use the calculator to bypass learning, to carry out procedures that you haven't learned and don't understand, the tool will not open up new horizons. It won't help you gain new mathematical knowledge and skills. It will simply be a black box, a mysterious number-producing mechanism. It will be a barrier to higher thought rather than a spur to it.

That's what computer automation often does today, and it's why Whitehead's observation has become misleading as a guide to technology's consequences. Rather than extending the brain's innate capacity for automaticity, automation too often becomes an impediment to automatization. In relieving us of repetitive mental exercise, it also relieves us of deep learning. Both complacency and bias are symptoms of a mind that is not being challenged, that is not fully engaged in the kind of real-world practice that generates knowledge, enriches memory, and builds skill. The problem is compounded by the way computer systems distance us from direct and immediate feedback about our actions. As the psychologist K. Anders Ericsson, an expert on talent development, points out, regular feedback is essential to skill building. It's what lets us learn from our mistakes and our successes. "In the absence of adequate feedback," Ericsson explains, "efficient learning is impossible and improvement only minimal even for highly motivated subjects."[31]

Automaticity, generation, flow: these mental phenomena are diverse, they're complicated, and their biological underpinnings are understood only fuzzily. But they are all related, and they tell us something important about ourselves. The kinds of effort that give rise to talent—characterized by challenging tasks, clear goals, and direct feedback—are very similar to those that provide us with a sense of flow. They're immersive experiences. They also describe the kinds of work that force us to actively generate knowledge rather than passively take in information. Honing our skills, enlarging our understanding, and achieving personal satisfaction and fulfillment are all of a piece. And they all require tight connections, physical and mental, between the individual and the world. They all require, to quote the American philosopher Robert Talisse, "getting your hands dirty with the world and letting the world kick back in a certain way."[32] Automaticity is the inscription the world leaves on the active mind and the active self. Know-how is the evidence of the richness of that inscription.

From rock climbers to surgeons to pianists, Mihaly Csikszentmihalyi explains, people who "routinely find deep enjoyment in an activity illustrate how an organized set of challenges and a corresponding set of skills result in optimal experience." The jobs or hobbies they engage in "afford rich opportunities for action," while the skills they develop allow them to make the most of those opportunities. The ability to act with aplomb in the world turns all of us into artists. "The effortless absorption experienced by the practiced artist at work on a difficult project always is premised upon earlier mastery of a complex body of skills."[33] When automation distances us from our work, when it gets between us and the world, it erases the artistry from our lives.

Interlude,
with Dancing Mice

"SINCE 1903 I HAVE HAD UNDER OBSERVATION CONSTANTLY from two to one hundred dancing mice." So confessed the Harvard psychologist Robert M. Yerkes in the opening chapter of his 1907 book *The Dancing Mouse*, a 290-page paean to a rodent. But not just any rodent. The dancing mouse, Yerkes predicted, would prove as important to the behavioralist as the frog was to the anatomist.

When a local Cambridge doctor presented a pair of Japanese dancing mice to the Harvard Psychological Laboratory as a gift, Yerkes was underwhelmed. It seemed "an unimportant incident in the course of my scientific work." But in short order he became infatuated with the tiny creatures and their habit of "whirling around on the same spot with incredible rapidity." He bred scores of them, assigning each a number and keeping a meticulous log of its markings, gender, birth date, and ancestry. A "really admirable animal," the dancing mouse was, he wrote, smaller and weaker than the average mouse—it was barely able to hold itself upright or "cling to an object"—but it proved "an ideal subject for the experimental study of many of the problems of animal behavior." The breed was "easily cared for, readily tamed, harmless, incessantly active, and it lends itself satisfactorily to a large number of experimental situations."[1]

At the time, psychological research using animals was still new. Ivan Pavlov had only begun his experiments on salivating dogs in the 1890s, and it wasn't until 1900 that an American graduate student named Willard Small dropped a rat into a maze and watched it scurry about. With his dancing mice, Yerkes greatly expanded the scope of animal studies. As he catalogued in *The Dancing Mouse*, he used the rodents as test subjects in the exploration of, among other things, balance and equilibrium, vision and perception, learning and memory, and the inheritance of behavioral traits. The mice were "experiment-impelling," he reported. "The longer I observed and experimented with them, the more numerous became the problems which the dancers presented to me for solution."[2]

Early in 1906, Yerkes began what would turn out to be his most important and influential experiments on the dancers. Working with his student John Dillingham Dodson, he put, one by one, forty of the mice into a wooden box. At the far end of the box were two passageways, one painted white, the other black. If a mouse tried to enter the black passageway, it received, as Yerkes and Dodson later wrote, "a disagreeable electric shock." The intensity of the jolt varied. Some mice were given a weak shock, others were given a strong one, and still others were given a moderate one. The researchers wanted to see if the strength of the stimulus would influence the speed with which the mice learned to avoid the black passage and go into the white one. What they discovered surprised them. The mice receiving the weak shock were relatively slow to distinguish the white and the black passageways, as might be expected. But the mice receiving the strong shock exhibited equally slow learning. The rodents quickest to understand their situation and modify their behavior were the ones given a moderate shock. "Contrary to our expectations," the scientists reported, "this set of experiments did not prove that the rate of habit-formation increases with increase in the strength of the electric stimulus up to the point at which the shock becomes positively injurious. Instead an intermediate range of intensity of stimulation proved to be most favorable to the acquisition of a habit."[3]

A subsequent series of tests brought another surprise. The scientists put a new group of mice through the same drill, but this time they increased the brightness of the light in the white passageway and dimmed the light in the black one, strengthening the visual contrast between the two. Under this condition, the mice receiving the strongest shock were the quickest to avoid the black doorway. Learning didn't fall off as it had in the first go-round. Yerkes and Dodson traced the difference in the rodents' behavior to the fact that the setup of the second experiment had made things easier for the animals. Thanks to the greater visual contrast, the mice didn't have to think as hard in distinguishing the passageways and associating the shock with the dark corridor. "The relation of the strength of electrical stimulus to rapidity of learning or habit-formation depends upon the difficultness of the habit," they explained.[4] As a task becomes harder, the optimum amount of stimulation decreases. In other words, when the mice faced a really tough challenge, both an unusually weak stimulus and an unusually strong stimulus impeded their learning. In something of a Goldilocks effect, a moderate stimulus inspired the best performance.

Since its publication in 1908, the paper that Yerkes and Dodson wrote about their experiments, "The Relation of Strength of Stimulus to Rapidity of Habit-Formation," has come to be recognized as a landmark in the history of psychology. The phenomenon they discovered, known as the Yerkes-Dodson law, has been observed, in various forms, far beyond the world of dancing mice and differently colored doorways. It affects people as well as rodents. In its human manifestation, the law is usually depicted as a bell curve that plots the relation of a person's performance at a difficult task to the level of mental stimulation, or arousal, the person is experiencing.

At very low levels of stimulation, the person is so disengaged and uninspired as to be moribund; performance flat-lines. As stimulation picks up, performance strengthens, rising steadily along the left side of the bell curve until it reaches a peak. Then, as stimulation continues to intensify, performance drops off, descending steadily down the right side of the bell. When stimulation reaches its most intense level, the person essentially becomes paralyzed with stress; performance again flat-lines. Like dancing mice, we humans learn and perform best when we're at the peak of the Yerkes-Dodson curve, where we're challenged but not overwhelmed. At the top of the bell is where we enter the state of flow.

The Yerkes-Dodson law has turned out to have particular pertinence to the study of automation. It helps explain many of the unexpected consequences of introducing computers into work places and processes. In automation's early days, it was thought that software, by handling routine chores, would reduce people's workload and enhance their performance. The assumption was that workload and performance were inversely correlated. Ease a person's mental strain, and she'll be smarter and sharper on the job. The reality has turned out to be more complicated. Sometimes, computers succeed in moderating workload in a way that allows a person to excel at her work, devoting her full attention to the most pressing tasks. In other cases, automation ends up reducing workload too much. The worker's performance suffers as she drifts to the left side of the Yerkes-Dodson curve.

We all know about the ill effects of information overload. It turns out that information underload can be equally debilitating. However well intentioned, making things easy for people can backfire. Human-factors scholars Mark Young and Neville Stanton have found evidence that a person's "attentional capacity" actually "shrinks to accommodate reductions in mental workload." In the operation of automated systems, they argue, "underload is possibly of greater concern [than overload], as it is more difficult to detect."[5] Researchers worry that the lassitude produced by information underload is going to be a particular danger with coming generations of automotive automation. As software takes over more steering and braking chores, the person behind the wheel won't have enough to do and will tune out. Making matters worse, the driver will likely have received little or no training in the use and risks of automation. Some routine accidents may be avoided, but we're going to end up with even more bad drivers on the road.

In the worst cases, automation actually places added and unexpected demands on people, burdening them with extra work and pushing them to the right side of the Yerkes-Dodson curve. Researchers refer to this as the "automation paradox." As Mark Scerbo, a human-factors expert at Virginia's Old Dominion University, explains, "The irony behind automation arises from a growing body of research demonstrating that automated systems often *increase* workload and create *unsafe* working conditions."[6] If, for example, the operator of a highly automated chemical plant is suddenly plunged into a fast-moving crisis, he may be overwhelmed by the need to monitor information displays and manipulate various computer controls while also following checklists, responding to alerts and alarms, and taking other emergency measures. Instead of relieving him of distractions and stress, computerization forces him to deal with all sorts of additional tasks and stimuli. Similar problems crop up during cockpit emergencies, when pilots are required to input data into their flight computers and scan information displays even as they're struggling to take manual control of the plane. Anyone who's gone off course while following directions from a mapping app knows firsthand how computer automation can cause sudden spikes in workload. It's not easy to fiddle with a smartphone while driving a car.

What we've learned is that automation has a sometimes-tragic tendency to increase the complexity of a job at the worst possible moment—when workers already have too much to handle. The computer, introduced as an aid to reduce the chances of human error, ends up making it more likely that people, like shocked mice, will make the wrong move.

CHAPTER FIVE

WHITE-COLLAR COMPUTER

Late in the summer of 2005, researchers at the venerable RAND Corporation in California made a stirring prediction about the future of American medicine. Having completed what they called "the most detailed analysis ever conducted of the potential benefits of electronic medical records," they declared that the U.S. health-care system "could save more than $81 billion annually and improve the quality of care" if hospitals and physicians automated their record keeping. The savings and other benefits, which RAND had estimated "using computer simulation models," made it clear, one of the think tank's top scientists said, "that it is time for the government and others who pay for health care to aggressively promote health information technology."[1] The last sentence in a subsequent report detailing the research underscored the sense of urgency: "The time to act is now."[2]

When the RAND study appeared, excitement about the computerization of medicine was already running high. Early in 2004, George W. Bush had issued a presidential order establishing the Health Information Technology Adoption Initiative with the goal of digitizing most U.S. medical records within ten years. By the end of 2004, the federal government was handing out millions of dollars in grants to encourage the purchase of automated systems by doctors and hospitals. In June of 2005, the Department of Health and Human Services established a task force of government officials and industry executives, the American Health Information Community, to help spur the adoption of electronic medical records. The RAND research, by putting the anticipated benefits of electronic records into hard and seemingly reliable numbers, stoked both the excitement and the spending. As the *New York Times* would later report, the study "helped drive explosive growth in the electronic records industry and encouraged the federal government to give billions of dollars in financial incentives to hospitals and doctors that put the systems in place."[3] Shortly after being sworn in as president in 2009, Barack Obama cited the RAND numbers when he announced a program to dole out an additional $30 billion in government funds to subsidize purchases of electronic medical record (EMR) systems. A frenzy of investment ensued, as some three hundred thousand doctors and four thousand hospitals availed themselves of Washington's largesse.[4]

Then, in 2013, just as Obama was being sworn in for a second term, RAND issued a new and very different report on the prospects for information technology in health care. The exuberance was gone; the tone now was chastened and apologetic. "Although the use of health IT has increased," the authors of the paper wrote, "quality and efficiency of patient care are only marginally better. Research on the effectiveness of health IT has yielded mixed results. Worse yet, annual aggregate expenditures on health care in the United States have grown from approximately $2 trillion in 2005 to roughly $2.8 trillion today." Worst of all, the EMR systems that doctors rushed to install with taxpayer money are plagued by problems with "interoperability." The systems can't talk to each other, which leaves critical patient data locked up in individual hospitals and doctors' offices. One of the great promises of health IT has always been that it would, as the RAND authors noted, allow "a patient or provider to access needed health information anywhere at any time," but because current EMR applications employ proprietary formats and conventions, they simply "enforce brand loyalty to a particular health care system." While RAND continued to express high hopes for the future, it confessed that the "rosy scenario" in its original report had not panned out.[5]

Other studies back up the latest RAND conclusions. Although EMR systems are becoming common in the United States, and have been common in other countries, such as the United Kingdom and Australia, for years, evidence of their benefits remains elusive. In a broad 2011 review, a team of British public-health researchers examined more than a hundred recently published studies of computerized medical systems. They concluded that when it comes to patient care and safety, there's "a vast gap between the theoretical and empirically demonstrated benefits." The research that has been used to promote the adoption of the systems, the scholars found, is "weak and inconsistent," and there is "insubstantial evidence to support the cost-effectiveness of these technologies." As for electronic medical records in particular, the investigators reported that the existing research is inconclusive and provides "only anecdotal evidence of the fundamental expected benefits and risks."[6] Some other researchers offer slightly sunnier assessments. Another 2011 literature review, by Department of Health and Human Services staffers, found that "a large majority of the recent studies show measurable benefits emerging from the adoption of health information technology." But noting the limitations of the existing research, they also concluded that "there is only suggestive evidence that more advanced systems or specific health IT components facilitate greater benefits."[7] To date, there is no strong empirical support for claims that automating medical record keeping will lead to major reductions in health-care costs or significant improvements in the well-being of patients.

But if doctors and patients have seen few benefits from the scramble to automate record keeping, the companies that supply the systems have profited. Cerner Corporation, a medical software outfit, saw its revenues triple, from $1 billion to $3 billion, between 2005 and 2013. Cerner, as it happens, was one of five corporations that provided RAND with funding for the original 2005 study. The other sponsors, which included General Electric and Hewlett Packard, also have substantial business interests in health-care automation. As today's flawed systems are replaced or upgraded in the future, to fix their interoperability problems and other shortcomings, information technology companies will reap further windfalls.

■   ■   ■   ■

THERE'S NOTHING unusual about this story. A rush to install new and untested computer systems, particularly when spurred by grand claims from technology companies and analysts, almost always produces great disappointments for the buyers and great profits for the sellers. That doesn't mean that the systems are doomed to be a bust. As bugs are worked out, features refined, and prices cut, even overhyped systems can eventually save companies a lot of money, not least by reducing their need to hire wage-earning workers. (The investments are, of course, far more likely to generate attractive returns when businesses are spending taxpayer money rather than their own.) This historical pattern seems likely to unfold again with EMR applications and related systems. As physicians and hospitals continue to computerize their record keeping and other operations— the generous government subsidies are still flowing—demonstrable efficiency gains may be achieved in some areas, and the quality of care may well improve for some patients, particularly when that care requires the coordinated efforts of several specialists. The fragmentation and cloistering of patient data are real problems in medicine, which well-designed, standardized information systems can help fix.

Beyond standing as yet another cautionary tale about rash investments in unproven software, the original RAND report, and the reaction to it, provide deeper lessons. For one thing, the projections of "computer simulation models" should always be viewed with skepticism. Simulations are also simplifications; they replicate the real world only imperfectly, and their outputs often reflect the biases of their creators. More important, the report and its aftermath reveal how deeply the substitution myth is entrenched in the way society perceives and evaluates automation. The RAND researchers assumed that beyond the obvious technical and training challenges in installing the systems, the shift from writing medical reports on paper to composing them with computers would be straightforward. Doctors, nurses, and other caregivers would substitute an automated method for a manual method, but they wouldn't significantly change how they practice medicine. In fact, studies show that computers can "profoundly alter patient care workflow processes," as a group of doctors and academics reported in the journal *Pediatrics* in 2006. "Although the intent of computerization is to improve patient care by making it safer and more efficient, the adverse effects and unintended consequences of workflow disruption may make the situation far worse."[8]

Falling victim to the substitution myth, the RAND researchers did not sufficiently account for the possibility that electronic records would have ill effects along with beneficial ones—a problem that plagues many forecasts about the consequences of automation. The overly optimistic analysis led to overly optimistic policy. As the physicians and medical professors Jerome Groopman and Pamela Hartzband noted in a withering critique of the Obama administration's subsidies, the 2005 RAND report "essentially ignore[d] downsides to electronic medical records" and also discounted earlier research that failed to find benefits in shifting from paper to digital records.[9] RAND's assumption that automation would be a substitute for manual work proved false, as human-factors experts would have predicted. But the damage, in wasted taxpayer money and misguided software installations, was done.

EMR systems are used for more than taking and sharing notes. Most of them incorporate decision-support software that, through on-screen checklists and prompts, provides guidance and suggestions to doctors during the course of consultations and examinations. The EMR information entered by the doctor then flows into the administrative systems of the medical practice or hospital, automating the generation of bills, prescriptions, test requests, and other forms and documents. One of the unexpected results is that physicians often end up billing patients for more and more costly services than they would have before the software was installed. As a doctor fills out a computer form during an examination, the system automatically recommends procedures—checking the eyes of a diabetes patient, say—that the doctor might want to consider performing. By clicking a checkbox to verify the completion of the procedure, the doctor not only adds a note to the record of the visit, but in many cases also triggers the billing system to add a new line item to the bill. The prompts may serve as useful reminders, and they may, in rare cases, prevent a doctor from overlooking a critical component of an exam. But they also inflate medical bills—a fact that system vendors have not been shy about highlighting in their sales pitches.[10]

Before doctors had software to prompt them, they were less likely to add an extra charge for certain minor procedures. The procedures were subsumed into more general charges—for an office visit, say, or a yearly physical. With the prompts, the individual charges get added to the invoice automatically. Just by making an action a little easier or a little more routine, the system alters the doctor's behavior in a small but meaningful way. The fact that the doctor often ends up making more money by following the software's lead provides a further incentive to defer to the system's judgment. Some experts worry that the monetary incentive may be a little too strong. In response to press reports about the unforeseen increase in medical charges resulting from electronic records, the federal government launched, in October 2012, an investigation to determine whether the new systems were abetting systematic overbilling or even outright fraud in the Medicare program. A 2014 report from the Office of the Inspector General warned that "health care providers can use [EMR] software features that may mask true authorship of the medical record and distort information in the record to inflate health care claims."[11]

There's also evidence that electronic records encourage doctors to order unnecessary tests, which also ends up increasing rather than reducing the cost of care. One study, published in the journal *Health Affairs* in 2012, showed that when doctors are able to easily call up a patient's past x-rays and other diagnostic images on a computer, they are more likely to order a new imaging test than if they lacked immediate access to the earlier images. Overall, doctors with computerized systems ordered new imaging tests in 18 percent of patient visits, while those without the systems ordered new tests in just 13 percent of visits. One of the common assumptions about electronic records is that by providing easy and immediate access to past test results, they would reduce the frequency of diagnostic testing. But this study indicates that, as its authors put it, "the reverse may be true." By making it so easy to receive and review test results, the automated systems appear to "provide subtle encouragement to physicians to order more imaging studies," the researchers argue. "In borderline situations, substituting a few keystrokes for the sometimes time-consuming task of tracking down results from an imaging facility may tip the balance in favor of ordering a test."[12] Here again we see how automation changes people's behavior, and the way work gets done, in ways that are virtually impossible to predict—and that may run directly counter to expectations.

■   ■   ■   ■

THE INTRODUCTION of automation into medicine, as with its introduction into aviation and other professions, has effects that go beyond efficiency and cost. We've already seen how software-generated highlights on mammograms alter, sometimes for better and sometimes for worse, the way radiologists read images. As physicians come to rely on computers to aid them in more facets of their everyday work, the technology is influencing the way they learn, the way they make decisions, and even their bedside manner.

A study of primary-care physicians who adopted electronic records, conducted by Timothy Hoff, a professor at SUNY's University at Albany School of Public Health, reveals evidence of what Hoff terms "deskilling outcomes," including "decreased clinical knowledge" and "increased stereotyping of patients." In 2007 and 2008, Hoff interviewed seventy-eight physicians from primary-care practices of various sizes in upstate New York. Three-fourths of the doctors were routinely using EMR systems, and most of them said they feared computerization was leading to less thorough, less personalized care. The physicians using computers told Hoff that they would regularly "cut-and-paste" boilerplate text into their reports on patient visits, whereas when they dictated notes or wrote them by hand they "gave greater consideration to the quality and uniqueness of the information being read into the record." Indeed, said the doctors, the very process of writing and dictation had served as a kind of "red flag" that forced them to slow down and "consider what they wanted to say." The doctors complained to Hoff that the homogenized text of electronic records can diminish the richness of their understanding of patients, undercutting their "ability to make informed decisions around diagnosis and treatment."[13]

Doctors' growing reliance on the recycling, or "cloning," of text is a natural outgrowth of the adoption of electronic records. EMR systems change the way clinicians take notes just as, years ago, the adoption of word-processing programs changed the way writers write and editors edit. The traditional practices of dictation and composition, whatever their benefits, come to feel slow and cumbersome when forced to compete with the ease and speed of cut-and-paste, drag-and-drop, and point-and-click. Stephen Levinson, a physician and the author of a standard textbook on medical record keeping and billing, sees extensive evidence of the rote reuse of old text in new records. As doctors employ computers to take notes on patients, he says, "records of every visit read almost word for word the same except for minor variations confined almost exclusively to the chief complaint." While such "cloned documentation" doesn't "make sense clinically" and "doesn't satisfy the patient's needs," it nevertheless becomes the default method simply because it is faster and more efficient—and, not least, because cloned text often incorporates lists of procedures that serve as another trigger for adding charges to patients' bills.[14]

What cloning shears away is nuance. Nearly all the contents of a typical electronic record "is boilerplate," one internist told Hoff. "The story's just not there. Not in my notes, not in other doctors' notes." The cost of diminished specificity and precision is compounded as cloned records circulate among other doctors. Physicians end up losing one of their main sources of on-the-job learning. The reading of dictated or handwritten notes from specialists has long provided an important educational benefit for primary-care doctors, deepening their understanding not only of individual patients but of everything from "disease treatments and their efficacy to new modes of diagnostic testing," Hoff writes. As those reports come to be composed more and more of recycled text, they lose their subtlety and originality, and they become much less valuable as learning tools.[15]

Danielle Ofri, an internist at Bellevue Hospital in New York City who has written several books on the practice of medicine, sees other subtle losses in the switch from paper to electronic records. Although flipping through the pages of a traditional medical chart may seem archaic and inefficient these days, it can provide a doctor with a quick but meaningful sense of a patient's health history, spanning many years. The more rigid way that computers present information actually tends to foreclose the long view. "In the computer," Ofri writes, "all visits look the same from the outside, so it is impossible to tell which were thorough visits with extensive evaluation and which were only brief visits for medication refills." Faced with the computer's relatively inflexible interface, doctors often end up scanning a patient's records for "only the last two or three visits; everything before that is effectively consigned to the electronic dust heap."[16]

A recent study of the shift from paper to electronic records at University of Washington teaching hospitals provides further evidence of how the format of electronic records can make it harder for doctors to navigate a patient's chart to find notes "of interest." With paper records, doctors could use the "characteristic penmanship" of different specialists to quickly home in on critical information. Electronic records, with their homogenized format, erase such subtle distinctions.[17] Beyond the navigational issues, Ofri worries that the organization of electronic records will alter the way physicians think: "The system encourages fragmented documentation, with different aspects of a patient's condition secreted in unconnected fields, so it's much harder to keep a global synthesis of the patient in mind."[18]

The automation of note taking also introduces what Harvard Medical School professor Beth Lown calls a "third party" into the exam room. In an insightful 2012 paper, written with her student Dayron Rodriquez, Lown tells of how the computer itself "competes with the patient for clinicians' attention, affects clinicians' capacity to be fully present, and alters the nature of communication, relationships, and physicians' sense of professional role."[19] Anyone who has been examined by a computer-tapping doctor probably has firsthand experience of at least some of what Lown describes, and researchers are finding empirical evidence that computers do indeed alter in meaningful ways the interactions between physician and patient. In a study conducted at a Veterans Administration clinic, patients who were examined by doctors taking electronic notes reported that "the computer adversely affected the amount of time the physician spent talking to, looking at, and examining them" and also tended to make the visit "feel less personal."[20] The clinic's doctors generally agreed with the patients' assessments. In another study, conducted at a large health maintenance organization in Israel, where the use of EMR systems is more common than in the United States, researchers found that during appointments with patients, primary-care physicians spend between 25 and 55 percent of their time looking at their computer screen. More than 90 percent of the Israeli doctors interviewed in the study said that electronic record keeping "disturbed communication with their patients."[21] Such a loss of focus is consistent with what psychologists have learned about how distracting it can be to operate a computer while performing some other task. "Paying attention to the computer and to the patient requires multitasking," observes Lown, and multitasking "is the opposite of mindful presence."[22]

The intrusiveness of the computer creates another problem that's been widely documented. EMR and related systems are set up to provide on-screen warnings to doctors, a feature that can help avoid dangerous oversights or mistakes. If, for instance, a physician prescribes a combination of drugs that could trigger an adverse reaction in a patient, the software will highlight the risk. Most of the alerts, though, turn out to be unnecessary. They're irrelevant, redundant, or just plain wrong. They seem to be generated not so much to protect the patient from harm as to protect the software vendor from lawsuits. (In bringing a third party into the exam room, the computer also brings in that party's commercial and legal interests.) Studies show that primary-care physicians routinely dismiss about nine out of ten of the alerts they receive. That breeds a condition known as *alert fatigue*. Treating the software as an electronic boy-who-cried-wolf, doctors begin to tune out the alerts altogether. They dismiss them so quickly when they pop up that even the occasional valid warning ends up being ignored. Not only do the alerts intrude on the doctor-patient relationship; they're served up in a way that can defeat their purpose.[23]

A medical exam or consultation involves an extraordinarily intricate and intimate form of personal communication. It requires, on the doctor's part, both an empathic sensitivity to words and body language and a coldly rational analysis of evidence. To decipher a complicated medical problem or complaint, a clinician has to listen carefully to a patient's story while at the same time guiding and filtering that story through established diagnostic frameworks. The key is to strike the right balance between grasping the specifics of the patient's situation and inferring general patterns and probabilities derived from reading and experience. Checklists and other decision guides can serve as valuable aids in this process. They bring order to complicated and sometimes chaotic circumstances. But as the surgeon and *New Yorker* writer Atul Gawande explained in his book *The Checklist Manifesto*, the "virtues of regimentation" don't negate the need for "courage, wits, and improvisation." The best clinicians will always be distinguished by their "expert audacity."[24] By requiring a doctor to follow templates and prompts too slavishly, computer automation can skew the dynamics of doctor-patient relations. It can streamline patient visits and bring useful information to bear, but it can also, as Lown writes, "narrow the scope of inquiry prematurely" and even, by provoking an automation bias that gives precedence to the screen over the patient, lead to misdiagnoses. Doctors can begin to display "'screen-driven' information-gathering behaviors, scrolling and asking questions as they appear on the computer rather than following the patient's narrative thread."[25]

Being led by the screen rather than the patient is particularly perilous for young practitioners, Lown suggests, as it forecloses opportunities to learn the most subtle and human aspects of the art of medicine—the tacit knowledge that can't be garnered from textbooks or software. It may also, in the long run, hinder doctors from developing the intuition that enables them to respond to emergencies and other unexpected events, when a patient's fate can be sealed in a matter of minutes. At such moments, doctors can't be methodical or deliberative; they can't spend time gathering and analyzing information or working through templates. A computer is of little help. Doctors have to make near-instantaneous decisions about diagnosis and treatment. They have to act. Cognitive scientists who have studied physicians' thought processes argue that expert clinicians don't use conscious reasoning, or formal sets of rules, in emergencies. Drawing on their knowledge and experience, they simply "see" what's wrong—oftentimes making a working diagnosis in a matter of seconds—and proceed to do what needs to be done. "The key cues to a patient's condition," explains Jerome Groopman in his book *How Doctors Think*, "coalesce into a pattern that the physician identifies as a specific disease or condition." This is talent of a very high order, where, Groopman says, "thinking is inseparable from acting."[26] Like other forms of mental automaticity, it develops only through continuing practice with direct, immediate feedback. Put a screen between doctor and patient, and you put distance between them. You make it much harder for automaticity and intuition to develop.

■   ■   ■   ■

IT DIDN'T take long, after their ragtag rebellion was crushed, for the surviving Luddites to see their fears come true. The making of textiles, along with the manufacture of many other goods, went from handicraft to industry within a few short years. The sites of production moved from homes and village workshops to large factories, which, to ensure access to sufficient laborers, materials, and customers, usually had to be built in or near cities. Craft workers followed the jobs, uprooting their families in a great wave of urbanization that was swollen by the loss of farming jobs to threshers and other agricultural equipment. Inside the new factories, ever more efficient and capable machines were installed, boosting productivity but also narrowing the responsibility and autonomy of those who operated the equipment. Skilled craftwork became unskilled factory labor.

Adam Smith had recognized how the specialization of factory jobs would lead to the deskilling of workers. "The man whose whole life is spent in performing a few simple operations, of which the effects too are, perhaps, always the same, or very nearly the same, has no occasion to exert his understanding, or to exercise his invention in finding out expedients for removing difficulties which never occur," he wrote in The Wealth of Nations. "He naturally loses, therefore, the habit of such exertion, and generally becomes as stupid and ignorant as it is possible for a human creature to become."[27] Smith viewed the degradation of skills as an unfortunate but unavoidable by-product of efficient factory production. In his famous example of the division of labor at a pin-manufacturing plant, the master pin-maker who once painstakingly crafted each pin is replaced by a squad of unskilled workers, each performing a narrow task: "One man draws out the wire, another straights it, a third cuts it, a fourth points it, a fifth grinds it at the top for receiving the head; to make the head requires two or three distinct operations; to put it on, is a peculiar business, to whiten the pins is another; it is even a trade by itself to put them into the paper; and the important business of making a pin is, in this manner, divided into about eighteen distinct operations."[28] None of the men knows how to make an entire pin, but working together, each plying his own peculiar business, they churn out far more pins than could an equal number of master craftsmen working separately. And because the workers require little talent or training, the manu-facturer can draw from a large pool of potential laborers, obviating the need to pay a premium for expertise.

Smith also appreciated how the division of labor eased the way for mechanization, which served to narrow workers' skills even fur-ther. Once a manufacturer had broken an intricate process into a series of well-defined "simple operations," it became relatively easy to design a machine to carry out each operation. The division of labor within a factory provided a set of specifications for its machinery. By the early years of the twentieth century, the deskilling of factory workers had become an explicit goal of industry, thanks to Frederick Winslow Taylor's philosophy of "scientific management." Believing, in line with Smith, that "the greatest prosperity" would be achieved "only when the work of [companies] is done with the smallest com-bined expenditure of human effort," Taylor counseled factory owners to prepare strict instructions for how each employee should use each machine, scripting every movement of the worker's body and mind.[29] The great flaw in traditional ways of working, Taylor believed, was that they granted too much initiative and leeway to individuals. Opti-mum efficiency could be achieved only through the standardization of work, enforced by "rules, laws, and formulae" and reflected in the very design of machines.[30]

Viewed as a system, the mechanized factory, in which worker and machine merge into a tightly controlled, perfectly productive unit, was a triumph of engineering and efficiency. For the individuals who became its cogs, it brought, as the Luddites had foreseen, a sacrifice not only of skill but of independence. The loss in autonomy was more than economic. It was existential, as Hannah Arendt would emphasize in her 1958 book *The Human Condition*: "Unlike the tools of workmanship, which at every given moment in the work process remain the servants of the hand, the machines demand that the laborer serve them, that he adjust the natural rhythm of his body to their mechanical movement."[31] Technology had progressed—if that's the right word—from simple tools that broadened the worker's latitude to complex machines that constrained it.

In the second half of the last century, the relation between worker and machine grew more complicated. As companies expanded, technological progress accelerated, and consumer spending exploded, employment branched out into new forms. Managerial, professional, and clerical positions proliferated, as did jobs in the service sector. Machines assumed a welter of new forms as well, and people used them in all sorts of ways, on the job and off. The Taylorist ethos of achieving efficiency through the standardization of work processes, though still exerting a strong influence on business operations, was tempered in some companies by a desire to tap workers' ingenuity and creativity. The coglike employee was no longer the ideal. Brought into this situation, the computer quickly took on a dual role. It served a Taylorist function of monitoring, measuring, and controlling people's work; companies found that software applications provided a powerful means for standardizing processes and preventing deviations. But in the form of the PC, the computer also became a flexible, personal tool that granted individuals greater initiative and autonomy. The computer was both enforcer and emancipator.

As the uses of automation multiplied and spread from factory to office, the strength of the connection between technological progress and the deskilling of labor became a topic of fierce debate among sociologists and economists. In 1974, the controversy came to a head when Harry Braverman, a social theorist and onetime coppersmith, published a passionate book with a dry title, *Labor and Monopoly Capital: The Degradation of Work in the Twentieth Century*. In reviewing recent trends in employment and workplace technology, Braverman argued that most workers were being funneled into routine jobs that offered little responsibility, little challenge, and little opportunity to gain know-how in anything important. They often acted as accessories to their machines and computers. "With the development of the capitalist mode of production," he wrote, "the very concept of skill becomes degraded along with the degradation of labor, and the yardstick by which it is measured shrinks to such a point that today the worker is considered to possess a 'skill' if his or her job requires a few days' or weeks' training, several months of training is regarded as unusually demanding, and the job that calls for a learning period of six months or a year—such as computer programming—inspires a paroxysm of awe."[32] The typical craft apprenticeship, he pointed out, by way of comparison, had lasted at least four years and often as many as seven. Braverman's dense, carefully argued treatise was widely read. Its Marxist perspective fit with the radical atmosphere of the 1960s and early 1970s as neatly as a tenon in a mortise.

Braverman's argument didn't impress everyone.[33] Critics of his work—and there were plenty—accused him of overstating the importance of traditional craft workers, who even in the eighteenth and nineteenth centuries hadn't accounted for all that large a proportion of the labor force. They also thought he placed too much value on the manual skills associated with blue-collar production jobs at the expense of the interpersonal and analytical skills that come to the fore in many white-collar and service posts. The latter criticism pointed to a bigger problem, one that complicates any attempt to diagnose and interpret broad shifts in skill levels across the economy.

Skill is a squishy concept. Talent can take many forms, and there's no good, objective way to measure or compare them. Is an eighteenth-century cobbler making a pair of shoes at a bench in his workshop more or less skilled than a twenty-first-century marketer using her computer to develop a promotional plan for a product? Is a plasterer more or less skilled than a hairdresser? If a pipefitter in a shipyard loses his job and, after some training, finds new work repairing computers, has he gone up or down the skill ladder? The criteria necessary to provide good answers to such questions elude us. As a result, debates about trends in deskilling, not to mention upskilling, reskilling, and other varieties of skilling, often bog down in bickering over value judgments.

But if the broad skill-shift theories of Braverman and others are fated to remain controversial, the picture becomes clearer when the focus shifts to particular trades and professions. In case after case, we've seen that as machines become more sophisticated, the work left to people becomes less so. Although it's now been largely forgotten, one of the most rigorous explorations of the effect of automation on skills was completed during the 1950s by the Harvard Business School professor James Bright. He examined, in exhaustive detail, the consequences of automation on workers in thirteen different industrial settings, ranging from an engine-manufacturing plant to a bakery to a feed mill. From the case studies, he derived an elaborate hierarchy of automation. It begins with the use of simple hand tools and proceeds up through seventeen levels to the use of complex machines programmed to regulate their own operation with sensors, feedback loops, and electronic controls. Bright analyzed how various skill requirements—physical effort, mental effort, dexterity, conceptual understanding, and so on—change as machines become more fully automated. He found that skill demands increase only in the very earliest stages of automation, with the introduction of power hand tools. As more complex machines are introduced, skill demands begin to slacken, and the demands ultimately fall off sharply when workers begin to use highly automated, self-regulating machinery. "It seems," Bright wrote in his 1958 book *Automation and Management*, "that the more automatic the machine, the less the operator has to do."[34]

To illustrate how deskilling proceeds, Bright used the example of a metalworker. When the worker uses simple manual tools, such as files and shears, the main skill requirements are job knowledge, including in this case an appreciation of the qualities and uses of metal, and physical dexterity. When power hand tools are introduced, the job grows more complicated and the cost of errors is magnified. The worker is called on to display "new levels of dexterity and decision-making" as well as greater attentiveness. He becomes a "machinist." But when hand tools are replaced by mechanisms that perform a series of operations, such as milling machines that cut and grind blocks of metal into precise three-dimensional shapes, "attention, decision-making, and machine control responsibilities are partially or largely reduced" and "the technical knowledge requirement of machine functioning and adjustment is reduced tremendously." The machinist becomes a "machine operator." When mechanization becomes truly automatic—when machines are programmed to control themselves—the worker "contributes little or no physical or mental effort to the production activity." He doesn't even require much job knowledge, as that knowledge has effectively gone into the machine through its design and coding. His job, if it still exists, is reduced to "patrolling." The metalworker becomes "a sort of watchman, a monitor, a helper." He might best be thought of as "a liaison man between machine and operating management." Overall, concluded Bright, "the progressive effect of automation is first to relieve the operator of manual effort and then to relieve him of the need to apply continuous mental effort."[35]

When Bright began his study, the prevailing assumption, among

business executives, politicians, and academics alike, was that automated machinery would demand greater skills and training on the
part of workers. Bright discovered, to his surprise, that the opposite
was more often the case: "I was startled to find that the upgrading
effect had not occurred to anywhere near the extent that is often
assumed. On the contrary, there was more evidence that automation
had reduced the skill requirements of the operating work force." In
a 1966 report for a U.S. government commission on automation and
employment, Bright reviewed his original research and discussed
the technological developments that had occurred in the succeeding
years. The advance of automation, he noted, had continued apace,
propelled by the rapid deployment of mainframe computers in business and industry. The early evidence suggested that the broad adoption of computers would continue rather than reverse the deskilling
trend. "The lesson," he wrote, "should be increasingly clear—it is
not necessarily true that highly complex equipment requires skilled
operators. The 'skill' can be built into the machine."[36]

■   ■   ■   ■

IT MAY seem as though a factory worker operating a noisy industrial
machine has little in common with a highly educated professional
entering esoteric information through a touchscreen or keyboard in
a quiet office. But in both cases, we see a person sharing a job with
an automated system—with another party. And, as Bright's work and
subsequent studies of automation make clear, the sophistication of
the system, whether it operates mechanically or digitally, determines
how roles and responsibilities are divided and, in turn, the set of
skills each party is called upon to exercise. As more skills are built
into the machine, it assumes more control over the work, and the
worker's opportunity to engage in and develop deeper talents, such as
those involved in interpretation and judgment, dwindles. When auto-

mation reaches its highest level, when it takes command of the job, the worker, skillwise, has nowhere to go but down. The immediate product of the joint machine-human labor, it's important to emphasize, may be superior, according to measures of efficiency and even quality, but the human party's responsibility and agency are nonetheless curtailed. "What if the cost of machines that think is people who don't?" asked George Dyson, the technology historian, in 2008.[37] It's a question that gains salience as we continue to shift responsibility for analysis and decision making to our computers.

The expanding ability of decision-support systems to guide doctors' thoughts, and to take control of certain aspects of medical decision making, reflects recent and dramatic gains in computing. When doctors make diagnoses, they draw on their knowledge of a large body of specialized information, learned through years of rigorous education and apprenticeship as well as the ongoing study of medical journals and other relevant literature. Until recently, it was difficult, if not impossible, for computers to replicate such deep, specialized, and often tacit knowledge. But inexorable advances in processing speed, precipitous declines in data-storage and networking costs, and breakthroughs in artificial-intelligence methods such as natural language processing and pattern recognition have changed the equation. Computers have become much more adept at reviewing and interpreting vast amounts of text and other information. By spotting correlations in the data—traits or phenomena that tend to be found together or to occur simultaneously or sequentially—computers are often able to make accurate predictions, calculating, say, the probability that a patient displaying a set of symptoms has or will develop a particular disease or the odds that a patient with a certain disease will respond well to a particular drug or other treatment regimen.

Through machine-learning techniques like decision trees and neural networks, which dynamically model complex statistical relationships among phenomena, computers are also able to refine the way they make predictions as they process more data and receive feedback about the accuracy of earlier guesses.[38] The weightings they give different variables get more precise, and their calculations of probability better reflect what happens in the real world. Today's computers get smarter as they gain experience, just as people do. New "neuromorphic" microchips, which have machine-learning protocols hardwired into their circuitry, will boost computers' learning ability in coming years, some computer scientists believe. Machines will become more discerning. We may bristle at the idea that computers are "smart" or "intelligent," but the fact is that while they may lack the understanding, empathy, and insight of doctors, computers are able to replicate many of the judgments of doctors through the statistical analysis of large amounts of digital information—what's come to be known as "big data." Many of the old debates about the meaning of intelligence are being rendered moot by the brute number-crunching force of today's data-processing machines.

The diagnostic skills of computers will only get better. As more data about individual patients are collected and stored, in the form of electronic records, digitized images and test results, pharmacy transactions, and, in the not-too-distant future, readings from personal biological sensors and health-monitoring apps, computers will become more proficient at finding correlations and calculating probabilities at ever finer levels of detail. Templates and guidelines will become more comprehensive and elaborate. Given the current stress on achieving greater efficiency in health care, we're likely to see the Taylorist ethos of optimization and standardization take hold throughout the medical field. The already strong trend toward replacing personal clinical judgment with the statistical outputs of so-called evidence-based medicine will gain momentum. Doctors will face increasing pressure, if not outright managerial fiat, to cede more control over diagnoses and treatment decisions to software.

To put it into uncharitable but not inaccurate terms, many doctors may soon find themselves taking on the role of human sensors who collect information for a decision-making computer. The doctors will examine the patient and enter data into electronic forms, but the computer will take the lead in suggesting diagnoses and recommending therapies. Thanks to the steady escalation of computer automation through Bright's hierarchy, physicians seem destined to experience, at least in some areas of their practice, the same deskilling effect that was once restricted to factory hands.

They will not be alone. The incursion of computers into elite professional work is happening everywhere. We've already seen how the thinking of corporate auditors is being shaped by expert systems that make predictions about risks and other variables. Other financial professionals, from loan officers to investment managers, also depend on computer models to guide their decisions, and Wall Street is now largely under the control of correlation-sniffing computers and the quants who program them. The number of people employed as securities dealers and traders in New York City plummeted by a third, from 150,000 to 100,000, between 2000 and 2013, despite the fact that Wall Street firms were often posting record profits. The overriding goal of brokerage and investment banking firms is "automating the system and getting rid of the traders," one financial industry analyst explained to a Bloomberg reporter. As for the traders who remain, "all they do today is hit buttons on computer screens."[39]

That's true not only in the trading of simple stocks and bonds but also in the packaging and dealing of complex financial instruments. Ashwin Parameswaran, a technology analyst and former investment banker, notes that "banks have made a significant effort to reduce the amount of skill and know-how required to price and trade financial derivatives. Trading systems have been progressively modified so that as much knowledge as possible is embedded within the software."[40]

Predictive algorithms are even moving into the lofty realm of venture capitalism, where top investors have long prided themselves on having a good nose for business and innovation. Prominent venture-capital firms like the Ironstone Group and Google Ventures now use computers to sniff out patterns in records of entrepreneurial success, and they place their bets accordingly.

A similar trend is under way in the law. For years, attorneys have depended on computers to search legal databases and prepare documents. Recently, software has taken a more central role in law offices. The critical process of document discovery, in which, traditionally, junior lawyers and paralegals read through reams of correspondence, email messages, and notes in search of evidence, has been largely automated. Computers can parse thousands of pages of digitized documents in seconds. Using e-discovery software with language-analysis algorithms, the machines not only spot relevant words and phrases but also discern chains of events, relationships among people, and even personal emotions and motivations. A single computer can take over the work of dozens of well-paid professionals. Document-preparation software has also advanced. By filling out a simple checklist, a lawyer can assemble a complex contract in an hour or two—a job that once took days.

On the horizon are bigger changes. Legal software firms are beginning to develop statistical prediction algorithms that, by analyzing many thousands of past cases, can recommend trial strategies, such as the choice of a venue or the terms of a settlement offer, that carry high probabilities of success. Software will soon be able to make the kinds of judgments that up to now required the experience and insight of a senior litigator.[41] Lex Machina, a company started in 2010 by a group of Stanford law professors and computer scientists, offers a preview of what's coming. With a database covering some 150,000 intellectual property cases, it runs computer analyses that predict the outcomes of patent lawsuits under various scenarios, taking into account the court, the presiding judge and participating attorneys, the litigants, the outcomes of related cases, and other factors.

Predictive algorithms are also assuming more control over the decisions made by business executives. Companies are spending billions of dollars a year on "people analytics" software that automates decisions about hiring, pay, and promotion. Xerox now relies exclusively on computers to choose among applicants for its fifty thousand call-center jobs. Candidates sit at a computer for a half-hour personality test, and the hiring software immediately gives them a score reflecting the likelihood that they'll perform well, show up for work reliably, and stick with the job. The company extends offers to those with high scores and sends low scorers on their way.[42] UPS uses predictive algorithms to chart daily routes for its drivers. Retailers use them to determine the optimal arrangement of merchandise on store shelves. Marketers and ad agencies use them in deciding where and when to run advertisements and in generating promotional messages on social networks. Managers increasingly find themselves playing a subservient role to software. They review and rubber-stamp plans and decisions produced by computers.

There's an irony here. In shifting the center of the economy from physical goods to data flows, computers brought new status and wealth to information workers during the last decades of the twentieth century. People who made their living by manipulating signs and symbols on screens became the stars of the new economy, even as the factory jobs that had long buttressed the middle class were being transferred overseas or handed off to robots. The dot-com bubble of the late 1990s, when for a few euphoric years riches flooded out of computer networks and into personal brokerage accounts, seemed to herald the start of a golden age of unlimited economic opportunity— what technology boosters dubbed a "long boom." But the good times proved fleeting. Now we're seeing that, as Norbert Wiener predicted, automation doesn't play favorites. Computers are as good at analyzing symbols and otherwise parsing and managing information as they are at directing the moves of industrial robots. Even the people who operate complex computer systems are losing their jobs to software, as data centers, like factories, become increasingly automated. The vast server farms operated by companies like Google, Amazon, and Apple essentially run themselves. Thanks to virtualization, an engineering technique that uses software to replicate the functions of hardware components like servers, the facilities' operations can be monitored and controlled by algorithms. Network problems and application glitches can be detected and fixed automatically, often in a matter of seconds. It may turn out that the late twentieth century's "intellectualization of labor," as the Italian media scholar Franco Berardi has termed it,[43] was just a precursor to the early twenty-first century's automation of intellect.

It's always risky to speculate how far computers will go in mimicking the insights and judgments of people. Extrapolations based on recent computing trends have a way of turning into fantasies. But even if we assume, contrary to the extravagant promises of big-data evangelists, that there are limits to the applicability and usefulness of correlation-based predictions and other forms of statistical analysis, it seems clear that computers are a long way from bumping up against those limits. When, in early 2011, the IBM supercomputer Watson took the crown as the reigning champion of *Jeopardy!*, thrashing two of the quiz show's top players, we got a preview of where computers' analytical talents are heading. Watson's ability to decipher clues was astonishing, but by the standards of contemporary artificial-intelligence programming, the computer was not performing an exceptional feat. It was, essentially, searching a vast database of documents for potential answers and then, by working simultaneously through a variety of prediction routines, determining which answer had the highest probability of being the correct one. But it was performing that feat so quickly that it was able to outthink exceptionally smart people in a tricky test involving trivia, wordplay, and recall.

Watson represents the flowering of a new, pragmatic form of artificial intelligence. Back in the 1950s and 1960s, when digital computers were still new, many mathematicians and engineers, and quite a few psychologists and philosophers, came to believe that the human brain had to operate like some sort of digital calculating machine. They saw in the computer a metaphor and a model for the mind. Creating artificial intelligence, it followed, would be fairly straightforward: you'd figure out the algorithms that run inside our skulls and then you'd translate those programs into software code. It didn't work. The original artificial-intelligence strategy failed miserably. Whatever it is that goes on inside our brains, it turned out, can't be reduced to the computations that go on inside computers.* Today's computer scientists are taking a very different approach to artificial intelligence that's at once less ambitious and more effective. The goal is no longer to replicate the *process* of human thought—that's still beyond our ken—but rather to replicate its *results*. These scientists look at a particular product of the mind—a hiring decision, say, or an answer to a trivia question—and then program a computer to accomplish the same result in its own mindless way. The workings of Watson's circuits bear little resemblance to the workings of the mind of a person playing *Jeopardy!*, but Watson can still post a higher score.

In the 1930s, while working on his doctoral thesis, the British mathematician and computing pioneer Alan Turing came up with the idea of an "oracle machine." It was a kind of computer that, applying a set of explicit rules to a store of data through "some unspecified

---

* The use of terms like *neural network* and *neuromorphic processing* may give the impression that computers operate the way brains operate (or vice versa). But the terms shouldn't be taken literally; they're figures of speech. Since we don't yet know how brains operate, how thought and consciousness arise from the interplay of neurons, we can't build computers that work as brains do.

means," could answer questions that normally would require tacit human knowledge. Turing was curious to figure out "how far it is possible to eliminate intuition, and leave only ingenuity." For the purposes of his thought experiment, he posited that there would be no limit to the machine's number-crunching acumen, no upper bound to the speed of its calculations or the amount of data it could take into account. "We do not mind how much ingenuity is required," he wrote, "and therefore assume it to be available in unlimited supply."[44] Turing was, as usual, prescient. He understood, as few others did at the time, the latent intelligence of algorithms, and he foresaw how that intelligence would be released by speedy calculations. Computers and databases will always have limits, but in systems like Watson we see the arrival of operational oracle machines. What Turing could only imagine, engineers are now building. Ingenuity is replacing intuition.

Watson's data-analysis acumen is being put to practical use as a diagnostic aid for oncologists and other doctors, and IBM foresees further applications in such fields as law, finance, and education. Spy agencies like the CIA and the NSA are also reported to be testing the system. If Google's driverless car reveals the newfound power of computers to replicate our psychomotor skills, to match or exceed our ability to navigate the physical world, Watson demonstrates computers' newfound power to replicate our cognitive skills, to match or exceed our ability to navigate the world of symbols and ideas.

■   ■   ■   ■

BUT THE replication of the outputs of thinking is not thinking. As Turing himself stressed, algorithms will never replace intuition entirely. There will always be a place for "spontaneous judgments which are not the result of conscious trains of reasoning."[45] What really makes us smart is not our ability to pull facts from documents or decipher statistical patterns in arrays of data. It's our ability to make sense of things, to weave the knowledge we draw from observation and experience, from *living*, into a rich and fluid understanding of the world that we can then apply to any task or challenge. It's this supple quality of mind, spanning conscious and unconscious cognition, reason and inspiration, that allows human beings to think conceptually, critically, metaphorically, speculatively, wittily—to take leaps of logic and imagination.

Hector Levesque, a computer scientist and roboticist at the University of Toronto, provides an example of a simple question that people can answer in a snap but that baffles computers:

> The large ball crashed right through the table because it was made of Styrofoam. What was made of Styrofoam, the large ball or the table?

We come up with the answer effortlessly because we understand what Styrofoam is and what happens when you drop something on a table and what tables tend to be like and what the adjective *large* implies. We grasp the context, both of the situation and of the words used to describe it. A computer, lacking any true understanding of the world, finds the language of the question hopelessly ambiguous. It remains locked in its algorithms. Reducing intelligence to the statistical analysis of large data sets "can lead us," says Levesque, "to systems with very impressive performance that are nonetheless *idiot-savants*." They might be great at chess or *Jeopardy!* or facial recognition or other tightly circumscribed mental exercises, but they "are completely hopeless outside their area of expertise."[46] Their precision is remarkable, but it's often a symptom of the narrowness of their perception.

Even when aimed at questions amenable to probabilistic answers, computer analysis is not flawless. The speed and apparent exactitude of computer calculations can mask limitations and distortions in the underlying data, not to mention imperfections in the data-mining algorithms themselves. Any large data set holds an abundance of spurious correlations along with the reliable ones. It's not hard to be misled by mere coincidence or to conjure a phantom association.[47] Once a particular data set becomes the basis for important decisions, moreover, the data and its analysis become vulnerable to corruption. Seeking financial, political, or social advantage, people will try to game the system. As the social scientist Donald T. Campbell explained in a renowned 1976 paper, "The more any quantitative social indicator is used for social decision-making, the more subject it will be to corruption pressures and the more apt it will be to distort and corrupt the social processes it is intended to monitor."[48]

Flaws in data and algorithms can leave professionals, and the rest of us, susceptible to an especially pernicious form of automation bias. "The threat is that we will let ourselves be mindlessly bound by the output of our analyses even when we have reasonable grounds for suspecting something is amiss," warn Viktor Mayer-Schönberger and Kenneth Cukier in their 2013 book *Big Data*. "Or that we will attribute a degree of truth to the data which it does not deserve."[49] A particular risk with correlation-calculating algorithms stems from their reliance on data about the past to anticipate the future. In most cases, the future behaves as expected; it follows precedent. But on those peculiar occasions when conditions veer from established patterns, the algorithms can make wildly inaccurate predictions—a fact that has already spelled disaster for some highly computerized hedge funds and brokerage firms. For all their gifts, computers still display a frightening lack of common sense.

The more we embrace what Microsoft researcher Kate Crawford terms "data fundamentalism,"[50] the more tempted we'll be to devalue the many talents computers can't mimic—to grant so much control to software that we restrict people's ability to exercise the know-how that comes from real experience and that can often lead to creative, counterintuitive insights. As some of the unforeseen consequences of electronic medical records show, templates and formulas are necessarily reductive and can all too easily become straightjackets of the mind. The Vermont doctor and medical professor Lawrence Weed has, since the 1960s, been a forceful and eloquent advocate for using computers to help doctors make smart, informed decisions.[51] He's been called the father of electronic medical records. But even he warns that the current "misguided use of statistical knowledge" in medicine "systematically excludes the individualized knowledge and data essential to patient care."[52]

Gary Klein, a research psychologist who studies how people make decisions, has deeper worries. By forcing physicians to follow set rules, evidence-based medicine "can impede scientific progress," he writes. Should hospitals and insurers "mandate EBM, backed up by the threat of lawsuits if adverse outcomes are accompanied by any departure from best practices, physicians will become reluctant to try alternative treatment strategies that have not yet been evaluated using randomized controlled trials. Scientific advancement can become stifled if front-line physicians, who blend medical expertise with respect for research, are prevented from exploration and are discouraged from making discoveries."[53]

If we're not careful, the automation of mental labor, by changing the nature and focus of intellectual endeavor, may end up eroding one of the foundations of culture itself: our desire to understand the world. Predictive algorithms may be supernaturally skilled at discovering correlations, but they're indifferent to the underlying causes of traits and phenomena. Yet it's the deciphering of causation—the meticulous untangling of how and why things work the way they do—that extends the reach of human understanding and ultimately gives meaning to our search for knowledge. If we come to see automated calculations of probability as sufficient for our professional and social purposes, we risk losing or at least weakening our desire and motivation to seek explanations, to venture down the circuitous paths that lead toward wisdom and wonder. Why bother, if a computer can spit out "the answer" in a millisecond or two?

In his 1947 essay "Rationalism in Politics," the British philosopher Michael Oakeshott provided a vivid description of the modern rationalist: "His mind has no atmosphere, no changes of season and temperature; his intellectual processes, so far as possible, are insulated from all external influence and go on in the void." The rationalist has no concern for culture or history; he neither cultivates nor displays a personal perspective. His thinking is notable only for "the rapidity with which he reduces the tangle and variety of experience" into "a formula."[54] Oakeshott's words also provide us with a perfect description of computer intelligence: eminently practical and productive and entirely lacking in curiosity, imagination, and worldliness.

CHAPTER SIX

WORLD AND SCREEN

THE SMALL ISLAND OF IGLOOLIK, LYING OFF THE COAST of the Melville Peninsula in the Nunavut territory of the Canadian North, is a bewildering place in the winter. The average temperature hovers around twenty degrees below zero. Thick sheets of sea ice cover the surrounding waters. The sun is absent. Despite the brutal conditions, Inuit hunters have for some four thousand years ventured out from their homes on the island and traversed miles of ice and tundra in search of caribou and other game. The hunters' ability to navigate vast stretches of barren Arctic terrain, where landmarks are few, snow formations are in constant flux, and trails disappear overnight, has amazed voyagers and scientists ever since 1822, when the English explorer William Edward Parry noted in his journal the "astonishing precision" of his Inuit guide's geographic knowledge.[1] The Inuit's extraordinary wayfinding skills are born not of technological prowess—they've eschewed maps, compasses, and other instruments—but of a profound understanding of winds, snowdrift patterns, animal behavior, stars, tides, and currents. The Inuit are masters of perception.

Or at least they used to be. Something changed in Inuit culture

at the turn of the millennium. In the year 2000, the U.S. government lifted many of the restrictions on the civilian use of the global
positioning system. The accuracy of GPS devices improved even as
their prices dropped. The Igloolik hunters, who had already swapped
their dogsleds for snowmobiles, began to rely on computer-generated
maps and directions to get around. Younger Inuit were particularly
eager to use the new technology. In the past, a young hunter had to
endure a long and arduous apprenticeship with his elders, developing his wayfinding talents over many years. By purchasing a cheap
GPS receiver, he could skip the training and offload responsibility for
navigation to the device. And he could travel out in some conditions,
such as dense fog, that used to make hunting trips impossible. The
ease, convenience, and precision of automated navigation made the
Inuit's traditional techniques seem antiquated and cumbersome by
comparison.

But as GPS devices proliferated on Igloolik, reports began to
spread of serious accidents during hunts, some resulting in injuries
and even deaths. The cause was often traced to an overreliance on
satellites. When a receiver breaks or its batteries freeze, a hunter
who hasn't developed strong wayfinding skills can easily become lost
in the featureless waste and fall victim to exposure. Even when the
devices operate properly, they present hazards. The routes so meticulously plotted on satellite maps can give hunters a form of tunnel
vision. Trusting the GPS instructions, they'll speed onto dangerously
thin ice, over cliffs, or into other environmental perils that a skilled
navigator would have had the sense and foresight to avoid. Some of
these problems may eventually be mitigated by improvements in navigational devices or by better instruction in their use. What won't be
mitigated is the loss of what one tribal elder describes as "the wisdom
and knowledge of the Inuit."[2]

The anthropologist Claudio Aporta, of Carleton University in
Ottawa, has been studying Inuit hunters for years. He reports that

while satellite navigation offers attractive advantages, its adoption has already brought a deterioration in wayfinding abilities and, more generally, a weakened feel for the land. As a hunter on a GPS-equipped snowmobile devotes his attention to the instructions coming from the computer, he loses sight of his surroundings. He travels "blindfolded," as Aporta puts it.[3] A singular talent that has defined and distinguished a people for thousands of years may well evaporate over the course of a generation or two.

■   ■   ■   ■

THE WORLD is a strange, changeable, and dangerous place. Getting around in it demands of any animal a great deal of effort, mental and physical. For ages, human beings have been creating tools to reduce the strain of travel. History is, among other things, a record of the discovery of ingenious new ways to ease our passage through our environs, to make it possible to cross greater and more daunting distances without getting lost, roughed up, or eaten. Simple maps and trail markers came first, then star maps and nautical charts and terrestrial globes, then instruments like sounding weights, quadrants, astrolabes, compasses, octants and sextants, telescopes, hourglasses, and chronometers. Lighthouses were erected along shorelines, buoys set in coastal waters. Roads were paved, signs posted, highways linked and numbered. It has, for most of us, been a long time since we've had to rely on our wits to get around.

GPS receivers and other automated mapping and direction-plotting devices are the latest additions to our navigational toolkit. They also give the old story a new and worrisome twist. Earlier navigational aids, particularly those available and affordable to ordinary folks, were just that: aids. They were designed to give travelers a greater awareness of the world around them—to sharpen their sense of direction, provide them with advance warning of danger, highlight nearby landmarks and other points of orientation, and in general help them situate themselves in both familiar and alien settings. Satellite navigation systems can do all those things, and more, but they're not designed to deepen our involvement with our surroundings. They're designed to relieve us of the need for such involvement. By taking control of the mechanics of navigation and reducing our own role to following routine commands—turn left in five hundred yards, take the next exit, stay right, destination ahead—the systems, whether running through a dashboard, a smartphone, or a dedicated GPS receiver, end up isolating us from the environment. As a team of Cornell University researchers put it in a 2008 paper, "With the GPS you no longer need to know where you are and where your destination is, attend to physical landmarks along the way, or get assistance from other people in the car and outside of it." The automation of wayfinding serves to "inhibit the process of experiencing the physical world by navigation through it."[4]

As is so often the case with gadgets and services that ease our way through life, we've celebrated the arrival of inexpensive GPS units. The *New York Times* writer David Brooks spoke for many when, in a 2007 op-ed titled "The Outsourced Brain," he raved about the navigation system installed in his new car: "I quickly established a romantic attachment to my GPS. I found comfort in her tranquil and slightly Anglophilic voice. I felt warm and safe following her thin blue line." His "GPS goddess" had "liberated" him from the age-old "drudgery" of navigation. And yet, he grudgingly confessed, the emancipation delivered by his in-dash muse came at a cost: "After a few weeks, it occurred to me that I could no longer get anywhere without her. Any trip slightly out of the ordinary had me typing the address into her system and then blissfully following her satellite-fed commands. I found that I was quickly shedding all vestiges of geographic knowledge." The price of convenience was, Brooks wrote, a loss of "autonomy."[5] The goddess was also a siren.

We want to see computer maps as interactive, high-tech versions of paper maps, but that's a mistaken assumption. It's yet another manifestation of the substitution myth. Traditional maps give us context. They provide us with an overview of an area and require us to figure out our current location and then plan or visualize the best route to our next stop. Yes, they require some work—good tools always do—but the mental effort aids our mind in creating its own cognitive map of an area. Map reading, research has shown, strengthens our sense of place and hones our navigational skills—in ways that can make it easier for us to get around even when we don't have a map at hand. We seem, without knowing it, to call on our subconscious memories of paper maps in orienting ourselves in a city or town and determining which way to head to arrive at our destination. In one revealing experiment, researchers found that people's navigational sense is actually sharpest when they're facing north—the same way maps point.[6] Paper maps don't just shepherd us from one place to the next; they teach us how to think about space.

The maps generated by satellite-linked computers are different. They usually provide meager spatial information and few navigational cues. Instead of requiring us to puzzle out where we are in an area, a GPS device simply sets us at the center of the map and then makes the world circulate around us. In this miniature parody of the pre-Copernican universe, we can get around without needing to know where we are, where we've been, or which direction we're heading. We just need an address or an intersection, the name of a building or a shop, to cue the device's calculations. Julia Frankenstein, a German cognitive psychologist who studies the mind's navigational sense, believes it's likely that "the more we rely on technology to find our way, the less we build up our cognitive maps." Because computer navigation systems provide only "bare-bones route information, without the spatial context of the whole area," she explains, our brains don't receive the raw material required to form rich memories of places. "Developing a cognitive map from this reduced information is a bit like trying to get an entire musical piece from a few notes."[7]

Other scientists agree. A British study found that drivers using paper maps developed stronger memories of routes and landmarks than did those relying on turn-by-turn instructions from satellite systems. After completing a trip, the map users were able to sketch more precise and detailed diagrams of their routes. The findings, reported the researchers, "provide strong evidence that the use of a vehicle navigation system will impact negatively on the formation of drivers' cognitive maps."[8] A study of drivers conducted at the University of Utah found evidence of "inattentional blindness" in GPS users, which impaired their "wayfinding performance" and their ability to form visual memories of their surroundings.[9] GPS-wielding pedestrians appear to suffer the same disabilities. In an experiment conducted in Japan, researchers had a group of people walk to a series of destinations in a city. Some of the subjects were given hand-held GPS devices; others used paper maps. The ones with the maps took more direct routes, had to pause less often, and formed clearer memories of where they'd been than did the ones with the gadgets. An earlier experiment, involving German pedestrians exploring a zoo, produced similar results.[10]

The artist and designer Sara Hendren, commenting on a trip she made to attend a conference in an unfamiliar city, summed up how easy it is to become dependent on computer maps today—and how such dependency can short-circuit the mind's navigational faculties and impede the development of a sense of place. "I realized that I was using my phone's map application, with spoken cues, to make the same short trip between my hotel and a conference center just five minutes away, several days in a row," she recalled. "I was really just willfully turning off the sphere of perception that I've relied on heavily most of my life: I made no attempt to remember landmarks and relationships and the look or feel of roads and such." She worries that by "outsourcing my multi-modal responsiveness and memory," she is "impoverishing my overall sensory experience."[11]

■   ■   ■   ■

AS TALES of discombobulated pilots, truck drivers, and hunters demonstrate, a loss of navigational acumen can have dire consequences. Most of us, in our daily routines of driving and walking and otherwise getting around, are unlikely to find ourselves in such perilous spots. Which raises the obvious question: *Who cares?* As long as we arrive at our destination, does it really matter whether we maintain our navigational sense or offload it to a machine? An Inuit elder on Igloolik may have good reason to bemoan the adoption of GPS technology as a cultural tragedy, but those of us living in lands crisscrossed by well-marked roads and furnished with gas stations, motels, and 7-Elevens long ago lost both the custom of and the capacity for prodigious feats of wayfinding. Our ability to perceive and interpret topography, especially in its natural state, is already much reduced. Paring it away further, or dispensing with it altogether, doesn't seem like such a big deal, particularly if in exchange we get an easier go of it.

But while we may no longer have much of a cultural stake in the conservation of our navigational prowess, we still have a personal stake in it. We are, after all, creatures of the earth. We're not abstract dots proceeding along thin blue lines on computer screens. We're real beings in real bodies in real places. Getting to know a place takes effort, but it ends in fulfillment and in knowledge. It provides a sense of personal accomplishment and autonomy, and it also provides a sense of belonging, a feeling of being at home in a place rather than passing through it. Whether practiced by a caribou hunter on an ice floe or a bargain hunter on an urban street, wayfinding opens a path from alienation to attachment. We may grimace when we hear people talk of "finding themselves," but the figure of speech, however vain and shopworn, acknowledges our deeply held sense that *who we are* is tangled up in *where we are*. We can't extract the self from its surroundings, at least not without leaving something important behind.

A GPS device, by allowing us to get from point A to point B with the least possible effort and nuisance, can make our lives easier, perhaps imbuing us, as David Brooks suggests, with a numb sort of bliss. But what it steals from us, when we turn to it too often, is the joy and satisfaction of apprehending the world around us—and of making that world a part of us. Tim Ingold, an anthropologist at the University of Aberdeen in Scotland, draws a distinction between two very different modes of travel: *wayfaring* and *transport*. Wayfaring, he explains, is "our most fundamental way of being in the world." Immersed in the landscape, attuned to its textures and features, the wayfarer enjoys "an experience of movement in which action and perception are intimately coupled." Wayfaring becomes "an ongoing process of growth and development, or self-renewal." Transport, on the other hand, is "essentially destination-oriented." It's not so much a process of discovery *"along* a way of life" as a mere "carrying *across,* from location to location, of people and goods in such a way as to leave their basic natures unaffected." In transport, the traveler doesn't actually move in any meaningful way. "Rather, he is moved, becoming a passenger in his own body."[12]

Wayfaring is messier and less efficient than transport, which is why it has become a target for automation. "If you have a mobile phone with Google Maps," says Michael Jones, an executive in Google's mapping division, "you can go anywhere on the planet and have confidence that we can give you directions to get to where you want to go safely and easily." As a result, he declares, "No human ever has to feel lost again."[13] That certainly sounds appealing, as if some basic problem in our existence had been solved forever. And it fits the Silicon Valley obsession with using software to rid people's lives of "friction." But the more you think about it, the more you realize that to never confront the possibility of getting lost is to live in a state of perpetual dislocation. If you never have to worry about not knowing where you are, then you never have to know where you are. It is also to live in a state of dependency, a ward of your phone and its apps.

Problems produce friction in our lives, but friction can act as a catalyst, pushing us to a fuller awareness and deeper understanding of our situation. "When we circumvent, by whatever means, the demand a place makes of us to find our way through it," the writer Ari Schulman observed in his 2011 *New Atlantis* essay "GPS and the End of the Road," we end up foreclosing "the best entry we have into inhabiting that place—and, by extension, to really *being* anywhere at all."[14]

We may foreclose other things as well. Neuroscientists have made a series of breakthroughs in understanding how the brain perceives and remembers space and location, and the discoveries underscore the elemental role that navigation plays in the workings of mind and memory. In a landmark study conducted at University College London in the early 1970s, John O'Keefe and Jonathan Dostrovsky monitored the brains of lab rats as the rodents moved about an enclosed area.[15] As a rat became familiar with the space, individual neurons in its hippocampus—a part of the brain that plays a central role in memory formation—would begin to fire every time the animal passed a certain spot. These location-keyed neurons, which the scientists dubbed "place cells" and which have since been found in the brains of other mammals, including humans, can be thought of as the signposts the brain uses to mark out a territory. Every time you enter a new place, whether a city square or the kitchen of a neighbor's house, the area is quickly mapped out with place cells. The cells, as O'Keefe has explained, appear to be activated by a variety of sensory signals, including visual, auditory, and tactile cues, "each of which can be perceived when the animal is in a particular part of the environment."[16]

More recently, in 2005, a team of Norwegian neuroscientists, led by the couple Edvard and May-Britt Moser, discovered a different set of neurons involved in charting, measuring, and navigating space, which they named "grid cells." Located in the entorhinal cortex, a region closely related to the hippocampus, the cells create in the brain a precise geographic grid of space, consisting of an array of regularly spaced, equilateral triangles. The Mosers compared the grid to a sheet of graph paper in the mind, on which an animal's location is traced as it moves about.[17] Whereas place cells map out specific locations, grid cells provide a more abstract map of space that remains the same wherever an animal goes, providing an inner sense of dead reckoning. (Grid cells have been found in the brains of several mammal species; recent experiments with brain-implanted electrodes indicate that humans have them too.[18]) Working in tandem, and drawing on signals from other neurons that monitor bodily direction and motion, place and grid cells act, in the words of the science writer James Gorman, "as a kind of built-in navigation system that is at the very heart of how animals know where they are, where they are going and where they have been."[19]

In addition to their role in navigation, the specialized cells appear to be involved more generally in the formation of memories, particularly memories of events and experiences. In fact, O'Keefe and the Mosers, as well as other scientists, have begun to theorize that the "mental travel" of memory is governed by the same brain systems that enable us to get around in the world. In a 2013 article in *Nature Neuroscience*, Edvard Moser and his colleague György Buzsáki provided extensive experimental evidence that "the neuronal mechanisms that evolved to define the spatial relationship among landmarks can also serve to embody associations among objects, events and other types of factual information." Out of such associations we weave the memories of our lives. It may well be that the brain's navigational sense—its ancient, intricate way of plotting and recording movement through space—is the evolutionary font of all memory.[20]

What's more than a little scary is what happens when that font goes dry. Our spatial sense tends to deteriorate as we get older, and in the worst cases we lose it altogether.[21] One of the earliest and most debilitating symptoms of dementia, including Alzheimer's disease, is hippocampal and entorhinal degeneration and the consequent loss of locational memory.[22] Victims begin to forget where they are. Véronique Bohbot, a research psychiatrist and memory expert at McGill University in Montreal, has conducted studies demonstrating that the way people exercise their navigational skills influences the functioning and even the size of the hippocampus—and may provide protection against the deterioration of memory.[23] The harder people work at building cognitive maps of space, the stronger their underlying memory circuits seem to become. They can actually grow gray matter in the hippocampus—a phenomenon documented in London cab drivers—in a way that's analogous to the building of muscle mass through physical exertion. But when they simply follow turn-by-turn instructions in "a robotic fashion," Bohbot warns, they don't "stimulate their hippocampus" and as a result may leave themselves more susceptible to memory loss.[24] Bohbot worries that, should the hippocampus begin to atrophy from a lack of use in navigation, the result could be a general loss of memory and a growing risk of dementia. "Society is geared in many ways toward shrinking the hippocampus," she told an interviewer. "In the next twenty years, I think we're going to see dementia occurring earlier and earlier."[25]

Even if we routinely use GPS devices when driving and walking outdoors, it's been suggested, we'll still have to rely on our own minds to get around when we're walking through buildings and other places that GPS signals can't reach. The mental exercise of indoor navigation, the theory goes, may help protect the functioning of our hippocampus and related neural circuits. While that argument may have been reassuring a few years ago, it is less so today. Hungry for more data on people's whereabouts and eager for more opportunities to distribute advertising and other messages keyed to their location, software and smartphone companies are rushing to extend the scope of their computer-mapping tools to indoor areas like airports, malls, and office buildings.

Google has already incorporated thousands of floor plans into its mapping services, and it has begun sending its Street View photographers into shops, offices, museums, and even monasteries to create detailed maps and panoramas of enclosed spaces. The company is also developing a technology, code-named Tango, that uses motion sensors and cameras in people's smartphones to generate three-dimensional maps of buildings and rooms. In early 2013, Apple acquired WiFiSlam, an indoor mapping company that had invented a way to use ambient WiFi and Bluetooth signals, rather than GPS transmissions, to pinpoint a person's location to within a few inches. Apple quickly incorporated the technology into the iBeacon feature now built into its iPhones and iPads. Scattered around stores and other spaces, iBeacon transmitters act as artificial place cells, activating whenever a person comes within range. They herald the onset of what *Wired* magazine calls "microlocation" tracking.[26]

Indoor mapping promises to ratchet up our dependence on computer navigation and further limit our opportunities for getting around on our own. Should personal head-up displays, such as Google Glass, come into wide use, we would always have easy and immediate access to turn-by-turn instructions. We'd receive, as Google's Michael Jones puts it, "a continuous stream of guidance," directing us everywhere we want to go.[27] Google and Mercedes-Benz are already collaborating on an app that will link a Glass headset to a driver's in-dash GPS unit, enabling what the carmaker calls "door-to-door navigation."[28] With the GPS goddess whispering in our ear, or beaming her signals onto our retinas, we'll rarely, if ever, have to exercise our mental mapping skills.

Bohbot and other researchers emphasize that more research needs to be done before we'll know for sure whether long-term use of GPS devices weakens memory and raises the risk of senility. But given all we've learned about the close links between navigation, the hippocampus, and memory, it is entirely plausible that avoiding the work of figuring out where we are and where we're going may have unforeseen and less-than-salubrious consequences. Because memory is what enables us not only to recall past events but to respond intelligently to present events and plan for future ones, any degradation in its functioning would tend to diminish the quality of our lives.

Through hundreds of thousands of years, evolution has fit our bodies and minds to the environment. We've been formed by being, to appropriate a couple of lines from the poet Wordsworth,

> Rolled round in earth's diurnal course,
> With rocks, and stones, and trees.

The automation of wayfinding distances us from the environment that shaped us. It encourages us to observe and manipulate symbols on screens rather than attend to real things in real places. The labors our obliging digital deities would have us see as mere drudgery may turn out to be vital to our fitness, happiness, and well-being. So *Who cares?* probably isn't the right question. What we should be asking ourselves is, *How far from the world do we want to retreat?*

■   ■   ■   ■

THAT'S A question the people who design buildings and public spaces have been grappling with for years. If aviators were the first professionals to experience the full force of computer automation, architects and other designers weren't far behind. In the early 1960s, a young computer engineer at MIT named Ivan Sutherland invented Sketchpad, a revolutionary software application for drawing and drafting that was the first program to employ a graphical user interface. Sketchpad set the stage for the development of computer-aided design, or CAD. After CAD programs were adapted to run on personal computers in the 1980s, design applications that automated the creation of two-dimensional drawings and three-dimensional models proliferated. The programs quickly became essential tools for architects, not to mention product designers, graphic artists, and civil engineers. By the start of the twenty-first century, as William J. Mitchell, the late dean of MIT's architecture school, observed, "architectural practice without CAD technology had become as unimaginable as writing without a word processor."[29] The new software tools changed, in ways that are still playing out today, the process, character, and style of design. The recent history of the architectural trade provides a view into automation's influence not only on spatial perception but on creative work.

Architecture is an elegant occupation. It combines the artist's pursuit of beauty with the craftsman's attentiveness to function, while also requiring a sensitivity to financial, technical, and other practical constraints. "Architecture is at the edge, between art and anthropology, between society and science, technology and history," explains the Italian architect Renzo Piano, designer of the Pompidou Center in Paris and the New York Times Building in Manhattan. "Sometimes it's humanistic and sometimes it's materialistic."[30] The work of an architect bridges the imaginative mind and the calculative mind, two ways of thinking that are often in tension, if not outright conflict. Since most of us spend most of our time in designed spaces—the constructed world at this point feels more natural to us than nature itself—architecture also exerts a deep if sometimes unappreciated influence over us, individually and collectively. Good architecture elevates life, while bad or mediocre architecture diminishes or cheapens it. Even small details like the size and placement of a window or an air vent can have a big effect on the aesthetics, usefulness, and efficiency of a building—and the comfort and mood of those inside it. "We shape our buildings," remarked Winston Churchill, "and afterwards our buildings shape us."[31]

While computer-generated plans can breed complacency when it comes to checking measurements, design software has in general made architecture firms more efficient. CAD systems have sped up and simplified the production of construction documents and made it easier for architects to share their plans with clients, engineers, contractors, and public officials. Manufacturers can now use architects' CAD files to program robots to fabricate building components, allowing for greater customization of materials while also cutting out time-consuming data-entry and review steps. The systems give architects a comprehensive view of a complex project, encompassing its floor plans, elevations, and materials as well as its various systems for heating and cooling, electricity, lighting, and plumbing. The ripple effects of changes in a design can be seen immediately, in a way that wasn't possible when plans took the form of a large stack of paper documents. Drawing on a computer's ability to incorporate all sorts of variables into its calculations, architects can estimate with precision the energy efficiency of their structures under many conditions, fulfilling a need of ever greater concern to the building trade and society in general. Detailed 3-D computer renderings and animations have also proved invaluable as a means for visualizing the exterior and interior of a building. Clients can be led on virtual walk-throughs and fly-throughs long before construction begins.

Beyond the practical benefits, the speed and precision of CAD calculations and visualizations have given architects and engineers the chance to experiment with new forms, shapes, and materials.

Buildings that once existed only in the imagination are now being built. Frank Gehry's Experience Music Project, a Seattle museum that looks like a collection of wax sculptures melting in the sun, would not exist were it not for computers. Although Gehry's original design took the form of a physical model, fashioned from wood and cardboard, translating the model's intricate, fluid shapes into construction plans could not be done by hand. It required a powerful CAD system—originally developed by the French firm Dassault to design jet aircraft—that could scan the model digitally and express its whimsy as a set of numbers. The materials for the building were so various and oddly shaped that their fabrication had to be automated too. The thousands of intricately fitted panels that form the museum's stainless-steel and aluminum facade were cut according to measurements calculated by the CAD program and fed directly into a computer-aided manufacturing system.

Gehry has long operated on architecture's technological frontier, but his practice of building models by hand is itself starting to seem archaic. As young architects have become more adept with computer drafting and modeling, CAD software has gone from a tool for turning designs into plans to a tool for producing the designs themselves. The increasingly popular technique of parametric design, which uses algorithms to establish formal relationships among different design elements, puts the computer's calculative power at the center of the creative process. Using spreadsheet-like forms or software scripts, an architect-programmer plugs a series of mathematical rules, or parameters, into a computer—a ratio of window size to floor area, say, or the vectors of a curved surface—and lets the machine output the design. In the most aggressive application of the technique, a building's form can be generated automatically by a set of algorithms rather than composed manually by the designer's hand.

As is often the case with new design techniques, parametric design has spawned a novel style of architecture called parametricism. Inspired by the geometric complexities of digital animation and the frenetic, aseptic collectivism of social networks, parametricism rejects the orderliness of classical architecture in favor of free-flowing assemblages of baroque, futuristic shapes. Some traditionalists view parametricism as a distasteful fad, dismissing its productions as, to quote New York architect Dino Marcantonio, "little more than the blobs that one can produce with minimal effort on the computer."[32] In a more temperate critique published in *The New Yorker*, the architecture writer Paul Goldberger observed that while the "swoops and bends and twists" of digital designs can be alluring, they "often seem disconnected from anything other than their own, computer-generated reality."[33] But some younger architects see parametricism, together with other forms of "computational design," as the defining architectural movement of our time, the center of energy in the profession. At the 2008 Architecture Biennale in Venice, Patrik Schumacher, a director of the influential Zaha Hadid firm in London, issued a "Parametricism Manifesto" in which he proclaimed that "parametricism is the great new style after modernism." Thanks to computers, he said, the structures of the built world will soon be composed of "radiating waves, laminal flows, and spiraling eddies," resembling "liquids in motion," and "swarms of buildings" will "drift across the landscape" in concert with "dynamic swarms of human bodies."[34]

Whether or not those harmonic swarms materialize, the controversy over parametric design brings into the open the soul searching that has been going on in architecture ever since CAD's arrival. From the start, the rush to adopt design software has been shadowed with doubt and trepidation. Many of the world's most respected architects and architecture teachers have warned that an overreliance on computers can narrow designers' perspectives and degrade their talent and creativity. Renzo Piano, for one, grants that computers have become "essential" to the practice of architecture, but he also fears that designers are shifting too much of their work to software. While automation allows an architect to generate precise and seemingly accomplished 3-D designs quickly, the very speed and exactitude of the machine may cut short the messy and painstaking process of exploration that gives rise to the most inspired and meaningful designs. The allure of the work as it appears on the screen may be an illusion. "You know," Piano says, "computers are getting so clever that they seem a bit like those pianos where you push a button and it plays the cha-cha and then a rumba. You may play very badly, but you feel like a great pianist. The same is true now in architecture. You may find yourself in the position where you feel like you're pushing buttons and able to build everything. But architecture is about thinking. It's about slowness in some way. You need time. The bad thing about computers is that they make everything run very fast."[35] The architect and critic Witold Rybczynski makes a similar point. While praising the great technological leaps that have transformed his profession over the years, he argues that "the fierce productivity of the computer carries a price—more time at the keyboard, less time thinking."[36]

■   ■   ■   ■

ARCHITECTS HAVE always thought of themselves as artists, and before the coming of CAD the wellspring of their art was the drawing. A freehand sketch is similar to a computer rendering in that it serves an obvious communication function. It provides an architect with a compelling visual medium for sharing a design idea with a client or a colleague. But the act of drawing is not just a way of expressing thought; it's a way of thinking. "I haven't got an imagination that can tell me what I've got without drawing it," says the modernist architect Richard MacCormac. "I use drawing as a process of criticism and discovery."[37] Sketching provides a bodily conduit between the abstract and the tangible. "Drawings are not just end products:

they are part of the thought process of architectural design," explains Michael Graves, the celebrated architect and product designer. "Drawings express the interaction of our minds, eyes and hands."[38] The philosopher Donald Schön may have put it best when he wrote that an architect holds a "reflective conversation" with his drawings, a conversation that is also, through its physicality, a dialogue with the actual materials of construction.[39] Through the back-and-forth, the give-and-take between hand and eye and mind, an idea takes form, a creative spark begins its slow migration from the imagination into the world.

Veteran architects' intuitive sense of the centrality of sketching to creative thinking is supported by studies of drawing's cognitive underpinnings and effects. Sketches on paper serve to expand the capacity of working memory, allowing an architect to keep in mind many different design options and variations. At the same time, the physical act of drawing, by demanding strong visual focus and deliberate muscle movements, aids in the forming of long-term memories. It helps the architect recall earlier sketches, and the ideas behind them, as he tries out new possibilities. "When I draw something, I remember it," explains Graves. "The drawing is a reminder of the idea that caused me to record it in the first place."[40] Drawing also allows the architect to shift quickly between different levels of detail and different degrees of abstraction, viewing a design from many angles simultaneously and weighing the implications of changes in details for the overall structure. Through drawing, writes the British design scholar Nigel Cross in his book *Designerly Ways of Knowing*, an architect not only progresses toward a final design but also hashes out the nature of the problem he's trying to solve: "We have seen that sketches incorporate not only drawings of tentative solution concepts but also numbers, symbols and texts, as the designer relates what he knows of the design problem to what is emerging as a solution. Sketching enables exploration of the problem space and the solution space to proceed together." In the hands of a talented architect, a sketchpad becomes, Cross concludes, "a kind of intelligence amplifier."[41]

Drawing might best be thought of as manual thinking. It is as much tactile as cerebral, as dependent on the hand as on the brain. The act of sketching appears to be a means of unlocking the mind's hidden stores of tacit knowledge, a mysterious process crucial to any act of artistic creation and difficult if not impossible to accomplish through conscious deliberation alone. "Design knowledge is knowing in action," Schön observed, and it "is mainly tacit." Designers "can best (or only) gain access to their knowledge in action by putting themselves into the mode of doing."[42] Designing with software on a computer screen is also a mode of doing, but it's a different mode. It emphasizes the more formal side of the work—thinking logically through a building's functional requirements and how various architectural elements might best be combined to achieve them. By diminishing the involvement of the hand, that "tool of tools," as Aristotle called it, the computer circumscribes the physicality of the task and narrows the architect's perceptual field. In place of the organic, corporeal figures that emerge from the tip of a pencil or a piece of charcoal, CAD software substitutes, Schön argued, "symbolic, procedural representations," which "are bound to be incomplete or inadequate in relation to the actual phenomena of design."[43] Just as the GPS screen deadens the Inuit hunter to the Arctic environment's faint but profuse sensory signals, the CAD screen restricts the architect's perception and appreciation of the materiality of his work. The world recedes.

In 2012, the Yale School of Architecture held a symposium called "Is Drawing Dead?" The stark title reflects a growing sense that the architect's freehand sketch is being rendered obsolete by the computer. The transition from sketchpad to screen entails, many architects believe, a loss of creativity, of adventurousness. Thanks to the precision and apparent completeness of screen renderings, a designer working at a computer has a tendency to lock in, visually and cognitively, on a design at an early stage. He bypasses much of the reflective and exploratory playfulness that springs from the tentativeness and ambiguity of sketching. Researchers term this phenomenon "premature fixation" and trace its cause to "the disincentive for design changes once a large amount of detail and interconnectedness is built too quickly into a CAD model."[44] The designer at the computer also tends to emphasize formal experimentation at the expense of expressiveness. By weakening an architect's "personal, emotional connection with the work," Michael Graves argues, CAD software produces designs that, "while complex and interesting in their own way," often "lack the emotional content of a design derived from hand."[45]

The distinguished Finnish architect Juhani Pallasmaa makes a related point in his eloquent 2009 book *The Thinking Hand*. He argues that the growing reliance on computers is making it harder for designers to imagine the human qualities of their buildings—to inhabit their works in progress in the way that people will ultimately inhabit the finished structures. Whereas hand-drawn sketches and handmade models have "the same flesh of physical materiality that the material object being designed and the architect himself embody," computer operations and images exist "in a mathematicised and abstracted immaterial world." Pallasmaa believes that "the false precision and apparent finiteness of the computer image" can stunt an architect's aesthetic sense, leading to technically stunning but emotionally sterile designs. In drawing with a pen or pencil, he writes, "the hand follows the outlines, shapes and patterns of the object," but when manipulating a simulated image with software, "the hand usually selects the lines from a given set of symbols that have no analogical—or, consequently, haptic or emotional—relation to the object."[46]

The controversy over the use of computers in design professions will go on, and each side will offer compelling evidence and persuasive arguments. Design software, too, will continue to advance, in ways that may address some of the limitations of existing digital tools. But whatever the future brings, the experience of architects and other designers makes clear that the computer is never a neutral tool. It influences, for better or worse, the way a person works and thinks. A software program follows a particular routine, which makes certain ways of working easier and others harder, and the user of the program adapts to the routine. The character and the goals of the work, as well as the standards by which it is judged, are shaped by the machine's capabilities. Whenever a designer or artisan (or anyone else, for that matter) becomes dependent on a program, she also assumes the preconceptions of the program's maker. In time, she comes to value what the software can do and dismiss as unimportant or irrelevant or simply unimaginable what it cannot. If she doesn't adapt, she risks being marginalized in her profession.

Beyond the specifications of the programming, simply transferring work from the world to the screen entails deep changes in perspective. Greater stress is placed on abstraction, less on materiality. Calculative power grows; sensory engagement fades. The precise and the explicit take precedence over the tentative and the ambiguous. E. J. Meade, a founder of Arch11, a small architecture firm in Boulder, Colorado, praises the efficiencies of design software, but he worries that popular programs like Revit and SketchUp are becoming too prescriptive. A designer need only type in the dimensions of a wall or floor or other surface, and with a click of a button the software generates all the details, automatically drawing each board or concrete block, each tile, all the supports, the insulation, the mortar, the texture of the plaster. Meade believes the way architects work and think is becoming homogenized as a result, and the buildings they design are becoming more predictable. "When you flipped through architecture journals in the 1980s," he told me, "you saw the hand of the individual architect." Today, what you tend to see is the functioning of the software: "You can read the operation of the technology in the final product."[47]

Like their counterparts in medicine, many veteran designers fear that the growing reliance on automated tools and routines is making it harder for students and younger professionals to learn the subtleties of their trade. Jacob Brillhart, an architecture professor at the University of Miami, believes that the easy shortcuts provided by programs like Revit are undermining "the apprenticeship process." Relying on software to fill in design details and specify materials "only breeds more banal, lazy and uneventful designs that are void of intellect, imagination and emotion." He also sees, again echoing the experience of doctors, a cut-and-paste culture emerging in his profession, with younger architects "pulling details, elevations, and wall sections off the office server from past projects and reassembling them."[48] The connection between doing and knowing is breaking down.

The danger looming over the creative trades is that designers and artists, dazzled by the computer's superhuman speed, precision, and efficiency, will eventually take it for granted that the automated way is the best way. They'll agree to the trade-offs that software imposes without considering them. They'll rush down the path of least resistance, even though a little resistance, a little friction, might have brought out the best in them.

■   ■   ■   ■

"To really know shoelaces," the political scientist and motorcycle mechanic Matthew Crawford has observed, "you have to tie shoes." That's a simple illustration of a deep truth that Crawford explores in his 2009 book *Shop Class as Soulcraft*: "If thinking is bound up with action, then the task of getting an adequate *grasp* on the world, intellectually, depends on our doing stuff in it."[49] Crawford draws on the work of the German philosopher Martin Heidegger, who argued that the deepest form of understanding available to us "is not mere perceptual cognition, but, rather, a handling, using, and taking care of things, which has its own kind of 'knowledge.'"[50]

We tend to talk about knowledge work as if it's something different from and even incompatible with manual labor—I confess to having said as much in earlier sections of this book—but the distinction is a smug and largely frivolous one. All work is knowledge work. The carpenter's mind is no less animated and engaged than the actuary's. The architect's accomplishments depend as much on the body and its senses as the hunter's do. What is true of other animals is true of us: the mind is not sealed in the skull but extends throughout the body. We think not only with our brain but also with our eyes and ears, nose and mouth, limbs and torso. And when we use tools to extend our grasp, we think with them as well. "Thinking, or knowledge-getting, is far from being the armchair thing it is often supposed to be," wrote the American philosopher and social reformer John Dewey in 1916. "Hands and feet, apparatus and appliances of all kinds are as much a part of it as changes in the brain."[51] To act is to think, and to think is to act.

Our desire to segregate the mind's cogitations from the body's exertions reflects the grip that Cartesian dualism still holds on us. When we think about thinking, we're quick to locate our mind, and hence our self, in the gray matter inside our skull and to see the rest of the body as a mechanical life-support system that keeps the neural circuits charged. More than a fancy of philosophers like Descartes and his predecessor Plato, this dualistic view of mind and body as operating in isolation from each other appears to be a side effect of consciousness itself. Even though the bulk of the mind's work goes on behind the scenes, in the shadows of the unconscious, we're aware only of the small but brightly lit window that the conscious mind opens for us. And our conscious mind tells us, insistently, that it's separate from the body.

According to UCLA psychology professor Matthew Lieberman, the illusion stems from the fact that when we contemplate our bodies, we draw on a different part of our brain than when we contemplate our minds. "When you think about your body and the actions of your body, you recruit a prefrontal and parietal region on the outer surface of your right hemisphere," he explains. "When you think about your mind you instead recruit different prefrontal and parietal regions in the middle of the brain, where the two hemispheres touch each other." When different areas in the brain process experiences, the conscious mind interprets those experiences as belonging to different categories. While the "hard-wired illusion" of mind-body dualism doesn't reflect actual "distinctions in nature," Lieberman emphasizes, it nevertheless has "immediate psychological reality for us."[52]

The more we learn about ourselves, the more we realize how misleading that particular "reality" is. One of the most interesting and illuminating areas of study in contemporary psychology and neuroscience involves what's called *embodied cognition*. Today's scientists and scholars are confirming John Dewey's insight of a century ago: Not only are brain and body composed of the same matter, but their workings are interwoven to a degree far beyond what we assume. The biological processes that constitute "thinking" emerge not just from neural computations in the skull but from the actions and sensory perceptions of the entire body. "For example," explains Andy Clark, a philosopher of mind at the University of Edinburgh who has written widely on embodied cognition, "there's good evidence that the physical gestures we make while we speak actually reduce the ongoing cognitive load on the brain, and that the biomechanics of the muscle and tendon systems of the legs hugely simplify the problem of controlled walking."[53] The retina, recent research shows, isn't a passive sensor sending raw data to the brain, as was once assumed; it actively shapes what we see. The eye has smarts of its own.[54] Even our conceptual musings appear to involve the body's systems for sensing and moving. When we think abstractly or meta- phorically about objects or phenomena in the world—tree branches, say, or gusts of wind—we mentally reenact, or simulate, our physi- cal experience of the things.[55] "For creatures like us," Clark argues, "body, world, and action" are "co-architects of that elusive thing that we call the mind."[56]

How cognitive functions are distributed among the brain, the sensory organs, and the rest of the body is still being studied and debated, and some of the more extravagant claims made by embodied- cognition advocates, such as the suggestion that the individual mind extends outside the body into the surrounding environment, remain controversial. What is clear is that we can no more separate our think- ing from our physical being than we can separate our physical being from the world that spawned it. "Nothing about human experience remains untouched by human embodiment," writes the philosopher Shaun Gallagher: "from the basic perceptual and emotional processes that are already at work in infancy, to a sophisticated interaction with other people; from the acquisition and creative use of language, to higher cognitive faculties involving judgment and metaphor; from the exercise of free will in intentional action, to the creation of cultural artifacts that provide for further human affordances."[57]

The idea of embodied cognition helps explain, as Gallagher sug- gests, the human race's prodigious facility for technology. Tuned to the surrounding environment, our bodies and brains are quick to bring tools and other artifacts into our thought processes—to treat things, neurologically, as parts of our selves. If you walk with a cane or work with a hammer or fight with a sword, your brain will incorpo- rate the tool into its neuronal map of your body. The nervous system's blending of body and object is not unique to humans. Monkeys use sticks to dig ants and termites from the ground, elephants use leafy branches to swat away biting flies, dolphins use bits of sponge to protect themselves from scrapes while digging for food on the ocean floor. But *Homo sapiens*'s superior aptitude for conscious reasoning and planning enables us to design ingenious tools and instruments for all sorts of purposes, extending our mental as well as our physical capacities. We have an ancient tendency toward what Clark terms "cognitive hybridization," the mixing of the biological and the technological, the internal and the external.[58]

The ease with which we make technology part of our selves can also lead us astray. We can grant power to our tools in ways that may not be in our best interest. One of the great ironies of our time is that even as scientists discover more about the essential roles that physical action and sensory perception play in the development of our thoughts, memories, and skills, we're spending less time acting in the world and more time living and working through the abstract medium of the computer screen. We're disembodying ourselves, imposing sensory constraints on our existence. With the general-purpose computer, we've managed, perversely enough, to devise a tool that steals from us the bodily joy of working with tools.

Our belief, intuitive but erroneous, that our intellect operates in isolation from our body leads us to discount the importance of involving ourselves with the world of things. That in turn makes it easy to assume that a computer—which to all appearances is an artificial brain, a "thinking machine"—is a sufficient and indeed superior tool for performing the work of the mind. Google's Michael Jones takes it as a given that "people are about 20 IQ points smarter now," thanks to his company's mapping tools and other online services.[59] Tricked by our own brains, we assume that we sacrifice nothing, or at least nothing essential, by relying on software scripts to travel from place to place or to design buildings or to engage in other sorts of thoughtful and inventive work. Worse yet, we remain oblivious to the fact that there are alternatives. We ignore the ways that software programs and automated systems might be reconfigured so as not to weaken our grasp on the world but to strengthen it. For, as human-factors researchers and other experts on automation have found, there are ways to break the glass cage without losing the many benefits computers grant us.

CHAPTER SEVEN

AUTOMATION
FOR THE PEOPLE

WHO NEEDS HUMANS, ANYWAY?

That question, in one rhetorical form or another, comes up frequently in discussions of automation. If computers are advancing so rapidly, and if people by comparison seem slow, clumsy, and error prone, why not build immaculately self-contained systems that perform flawlessly without any human oversight or intervention? Why not take the human factor out of the equation altogether? "We need to let robots take over," declared the technology theorist Kevin Kelly in a 2013 *Wired* cover story. He pointed to aviation as an example: "A computerized brain known as the autopilot can fly a 787 jet unaided, but irrationally we place human pilots in the cockpit to babysit the autopilot 'just in case.'"[1] The news that a person was driving the Google car that crashed in 2011 prompted a writer at a prominent technology blog to exclaim, "More robo-drivers!"[2] Commenting on the struggles of Chicago's public schools, *Wall Street Journal* writer Andy Kessler remarked, only half-jokingly, "Why not forget the teachers and issue all 404,151 students an iPad or Android tablet?"[3] In a 2012 essay, the respected Silicon Valley venture capitalist Vinod Khosla suggested that health care will be much improved when medical software—which he dubs "Doctor Algorithm"—goes from assisting primary-care physicians in making diagnoses to replacing the doctors entirely. "Eventually," he wrote, "we won't need the average doctor."[4] The cure for imperfect automation is total automation.

That's a seductive idea, but it's simplistic. Machines share the fallibility of their makers. Sooner or later, even the most advanced technology will break down, misfire, or, in the case of a computerized system, encounter a cluster of circumstances that its designers and programmers never anticipated and that leave its algorithms baffled. In early 2009, just a few weeks before the Continental Connection crash in Buffalo, a US Airways Airbus A320 lost all engine power after hitting a flock of Canada geese on takeoff from LaGuardia Airport in New York. Acting quickly and coolly, Captain Chesley Sullenberger and his first officer, Jeffrey Skiles, managed, in three harrowing minutes, to ditch the crippled jet safely in the Hudson River. All passengers and crew were evacuated. If the pilots hadn't been there to "babysit" the A320, a craft with state-of-the-art automation, the jet would have crashed and everyone on board would almost certainly have perished. For a passenger jet to have all its engines fail is rare. But it's not rare for pilots to rescue planes from mechanical malfunctions, autopilot glitches, rough weather, and other unexpected events. "Again and again," Germany's *Der Spiegel* reported in a 2009 feature on airline safety, the pilots of fly-by-wire planes "run into new, nasty surprises that none of the engineers had predicted."[5]

The same is true elsewhere. The mishap that occurred while a person was driving Google's Prius was widely reported in the press; what we don't hear much about are all the times the backup drivers in Google cars, and other automated test vehicles, have to take the wheel to perform maneuvers the computers can't handle. Google requires that people drive its cars manually when on most urban and residential streets, and any employee who wants to operate one of the vehicles has to complete rigorous training in emergency driving techniques.[6] Driverless cars aren't quite as driverless as they seem.

In medicine, caregivers often have to overrule misguided instructions or suggestions offered by clinical computers. Hospitals have found that while computerized drug-ordering systems alleviate some common errors in dispensing medication, they introduce new problems. A 2011 study at one hospital revealed that the incidence of duplicated medication orders actually increased after drug ordering was automated.[7] Diagnostic software is also far from perfect. Doctor Algorithm may well give you the right diagnosis and treatment most of the time, but if your particular set of symptoms doesn't fit the probability profile, you're going to be glad that Doctor Human was there in the examination room to review and overrule the computer's calculations.

As automation technologies become more complicated and more interconnected, with a welter of links and dependencies among software instructions, databases, network protocols, sensors, and mechanical parts, the potential sources of failure multiply. Systems become susceptible to what scientists call "cascading failures," in which a small malfunction in one component sets off a far-flung and catastrophic chain of breakdowns. Ours is a world of "interdependent networks," a group of physicists reported in a 2010 *Nature* article. "Diverse infrastructures such as water supply, transportation, fuel and power stations are coupled together" through electronic and other links, which ends up making all of them "extremely sensitive to random failure." That's true even when the connections are limited to exchanges of data.[8]

Vulnerabilities become harder to discern too. With the industrial machinery of the past, explains MIT computer scientist Nancy Leveson in her book *Engineering a Safer World*, "interactions among components could be thoroughly planned, understood, anticipated, and guarded against," and the overall design of a system could be tested exhaustively before it was put into everyday use. "Modern, high-tech systems no longer have these properties." They're less "intellectually manageable" than were their nuts-and-bolts predecessors.[9] All the parts may work flawlessly, but a small error or oversight in system design—a glitch that might be buried in hundreds of thousands of lines of software code—can still cause a major accident.

The dangers are compounded by the incredible speed at which computers can make decisions and trigger actions. That was demonstrated over the course of a hair-raising hour on the morning of August 1, 2012, when Wall Street's largest trading firm, Knight Capital Group, rolled out a new automated program for buying and selling shares. The cutting-edge software had a bug that went undetected during testing. The program immediately flooded exchanges with unauthorized and irrational orders, trading $2.6 million worth of stocks every second. In the forty-five minutes that passed before Knight's mathematicians and computer scientists were able to track the problem to its source and shut the offending program down, the software racked up $7 billion in errant trades. The company ended up losing almost half a billion dollars, putting it on the verge of bankruptcy. Within a week, a consortium of other Wall Street firms bailed Knight out to avoid yet another disaster in the financial industry.

Technology improves, of course, and bugs get fixed. Flawlessness, though, remains an ideal that can never be achieved. Even if a perfect automated system could be designed and built, it would still operate in an imperfect world. Autonomous cars don't drive the streets of utopia. Robots don't ply their trades in Elysian factories. Geese flock. Lightning strikes. The conviction that we can build an entirely self-sufficient, entirely reliable automated system is itself a manifestation of automation bias.

Unfortunately, that conviction is common not only among technol-

ogy pundits but also among engineers and software programmers—
the very people who design the systems. In a classic 1983 article
in the journal *Automatica*, Lisanne Bainbridge, an engineering psychologist at University College London, described a conundrum that
lies at the core of computer automation. Because designers often
assume that human beings are "unreliable and inefficient," at least
when compared to a computer, they strive to give them as small a
role as possible in the operation of systems. People end up functioning as mere monitors, passive watchers of screens.[10] That's a job
that humans, with our notoriously wandering minds, are particularly
bad at. Research on vigilance, dating back to studies of British radar
operators watching for German submarines during World War II,
shows that even highly motivated people can't keep their attention
focused on a display of relatively stable information for more than
about half an hour.[11] They get bored; they daydream; their concentration drifts. "This means," Bainbridge wrote, "that it is humanly
impossible to carry out the basic function of monitoring for unlikely
abnormalities."[12]

And because a person's skills "deteriorate when they are not used,"
she added, even an experienced system operator will eventually begin
to act like "an inexperienced one" if his main job consists of watching rather than acting. As his instincts and reflexes grow rusty from
disuse, he'll have trouble spotting and diagnosing problems, and his
responses will be slow and deliberate rather than quick and automatic. Combined with the loss of situational awareness, the degradation of know-how raises the odds that when something goes wrong,
as it sooner or later will, the operator will react ineptly. And once that
happens, system designers will work to place even greater limits on
the operator's role, taking him further out of the action and making
it more likely that he'll mess up in the future. The assumption that
the human being will be the weakest link in the system becomes
self-fulfilling.

■   ■   ■   ■

ERGONOMICS, THE art and science of fitting tools and workplaces to the people who use them, dates back at least to the Ancient Greeks. Hippocrates, in "On Things Relating to the Surgery," provides precise instructions for how operating rooms should be lit and furnished, how medical instruments should be arranged and handled, even how surgeons should dress. In the design of many Greek tools, we see evidence of an exquisite consideration of the ways an implement's form, weight, and balance affect a worker's productivity, stamina, and health. In early Asian civilizations, too, there are signs that the instruments of labor were carefully designed with the physical and psychological well-being of the worker in mind.[13]

It wasn't until the Second World War, though, that ergonomics began to emerge, together with its more theoretical cousin cybernetics, as a formal discipline. Many thousands of inexperienced soldiers and other recruits had to be entrusted with complicated and dangerous weapons and machinery, and there was little time for training. Awkward designs and confusing controls could no longer be tolerated. Thanks to trailblazing thinkers like Norbert Wiener and U.S. Air Force psychologists Paul Fitts and Alphonse Chapanis, military and industrial planners came to appreciate that human beings play as integral a role in the successful workings of a complex technological system as do the system's mechanical components and electronic regulators. You can't optimize a machine and then force the worker to adapt to it, in rigid Taylorist fashion; you have to design the machine to suit the worker.

Inspired at first by the war effort and then by the drive to incorporate computers into commerce, government, and science, a large and dedicated group of psychologists, physiologists, neurobiologists, engineers, sociologists, and designers began to devote their varied talents to studying the interactions of people and machines. Their focus may have been the battlefield and the factory, but their aspiration was deeply humanistic: to bring people and technology together in a productive, resilient, and safe symbiosis, a harmonious human-machine partnership that would get the best from both sides. If ours is an age of complex systems, then ergonomists are our metaphysicians.

At least they should be. All too often, discoveries and insights from the field of ergonomics, or, as it's now commonly known, human-factors engineering, are ignored or given short shrift. Concerns about the effects of computers and other machines on people's minds and bodies have routinely been trumped by the desire to achieve maximum efficiency, speed, and precision—or simply to turn as big a profit as possible. Software programmers receive little or no training in ergonomics, and they remain largely oblivious to relevant human-factors research. It doesn't help that engineers and computer scientists, with their strict focus on math and logic, have a natural antipathy toward the "softer" concerns of their counterparts in the human-factors field. A few years before his death in 2006, the ergonomics pioneer David Meister, recalling his own career, wrote that he and his colleagues "always worked against the odds so that anything that was accomplished was almost unexpected." The course of technological progress, he wistfully concluded, "is tied to the profit motive; consequently, it has little appreciation of the human."[14]

It wasn't always so. People first began thinking about technological progress as a force in history in the latter half of the eighteenth century, when the scientific discoveries of the Enlightenment began to be translated into the practical machinery of the Industrial Revolution. That was also, and not coincidentally, a time of political upheaval. The democratic, humanitarian ideals of the Enlightenment culminated in the revolutions in America and France, and those ideals also infused society's view of science and technology. Technical advances were valued—by intellectuals, if not always by workers—as means to political reform. Progress was defined in social terms, with technology playing a supporting role. Enlightenment thinkers such as Voltaire, Joseph Priestley, and Thomas Jefferson saw, in the words of the cultural historian Leo Marx, "the new sciences and technologies not as ends in themselves, but as instruments for carrying out a comprehensive transformation of society."

By the middle of the nineteenth century, however, the reformist view had, at least in the United States, been eclipsed by a new and very different concept of progress in which technology itself played the starring role. "With the further development of industrial capitalism," writes Marx, "Americans celebrated the advance of science and technology with increasing fervor, but they began to detach the idea from the goal of social and political liberation." Instead, they embraced "the now familiar view that innovations in science-based technologies are in themselves a sufficient and reliable basis for progress."[15] New technology, once valued as a means to a greater good, came to be revered as a good in itself.

It's hardly a surprise, then, that in our own time the capabilities of computers have, as Bainbridge suggested, determined the division of labor in complex automated systems. To boost productivity, reduce labor costs, and avoid human error—to further progress—you simply allocate control over as many activities as possible to software, and as software's capabilities advance, you extend the scope of its authority even further. The more technology, the better. The flesh-and-blood operators are left with responsibility only for those tasks that the designers can't figure out how to automate, such as watching for anomalies or providing an emergency backup in the event of a system failure. People are pushed further and further out of what engineers term "the loop"—the cycle of action, feedback, and decision making that controls a system's moment-by-moment operations.

Ergonomists call the prevailing approach *technology-centered automation*. Reflecting an almost religious faith in technology, and an equally fervent distrust of human beings, it substitutes misanthropic goals for humanistic ones. It turns the glib "who needs humans?" attitude of the technophilic dreamer into a design ethic. As the resulting machines and software tools make their way into workplaces and homes, they carry that misanthropic ideal into our lives. "Society," writes Donald Norman, a cognitive scientist and author of several influential books about product design, "has unwittingly fallen into a machine-centered orientation to life, one that emphasizes the needs of technology over those of people, thereby forcing people into a supporting role, one for which we are most unsuited. Worse, the machine-centered viewpoint compares people to machines and finds us wanting, incapable of precise, repetitive, accurate actions." Although it now "pervades society," this view warps our sense of ourselves. "It emphasizes tasks and activities that we should not be performing and ignores our primary skills and attributes—activities that are done poorly, if at all, by machines. When we take the machine-centered point of view, we judge things on artificial, mechanical merits."[16]

It's entirely logical that those with a mechanical bent would take a mechanical view of life. The impetus behind invention is often, as Norbert Wiener put it, "the desires of the gadgeteer to see the wheels go round."[17] And it's equally logical that such people would come to control the design and construction of the intricate systems and software programs that now govern or mediate society's workings. They're the ones who know the code. As society becomes ever more computerized, the programmer becomes its unacknowledged legislator. By defining the human factor as a peripheral concern, the technologist also removes the main impediment to the fulfillment of his desires; the unbridled pursuit of technological progress becomes self-justifying. To judge technology primarily on its technological merits is to give the gadgeteer carte blanche.

In addition to fitting the dominant ideology of progress, the bias to let technology guide decisions about automation has practical advantages. It greatly simplifies the work of the system builders. Engineers and programmers need only take into account what computers and machines can do. That allows them to narrow their focus and winnow a project's specifications. It relieves them of having to wrestle with the complexities, vagaries, and frailties of the human body and psyche. But however compelling as a design tactic, the simplicity of technology-centered automation is a mirage. Ignoring the human factor does not remove the human factor.

In a much-cited 1997 paper, "Automation Surprises," the human-factors experts Nadine Sarter, David Woods, and Charles Billings traced the origins of the technology-focused approach. They described how it grew out of and continues to reflect the "myths, false hopes, and misguided intentions associated with modern technology." The arrival of the computer, first as an analogue machine and then in its familiar digital form, encouraged engineers and industrialists to take an idealistic view of electronically controlled systems, to see them as a kind of cure-all for human inefficiency and fallibility. The order and cleanliness of computer operations and outputs seemed heaven-sent when contrasted with the earthly messiness of human affairs. "Automation technology," Sarter and her colleagues wrote, "was originally developed in hope of increasing the precision and economy of operations while, at the same time, reducing operator workload and training requirements. It was considered possible to create an autonomous system that required little if any human involvement and therefore reduced or eliminated the opportunity for human error." That belief led, again with pristine logic, to the further assumption that "automated systems could be designed without much consideration for the human element in the overall system."[18]

The desires and beliefs underpinning the dominant design approach, the authors continued, have proved naive and damaging. While automated systems have often enhanced the "precision and economy of operations," they have fallen short of expectations in other respects, and they have introduced a whole new set of problems. Most of the shortcomings stem from "the fact that even highly automated systems still require operator involvement and therefore communication and coordination between human and machine." But because the systems have been designed without sufficient regard for the people who operate them, their communication and coordination capabilities are feeble. In consequence, the computerized systems lack the "complete knowledge" of the work and the "comprehensive access to the outside world" that only people can provide. "Automated systems do not know when to initiate communication with the human about their intentions and activities or when to request additional information from the human. They do not always provide adequate feedback to the human who, in turn, has difficulties tracking automation status and behavior and realizing there is a need to intervene to avoid undesirable actions by the automation." Many of the problems that bedevil automated systems stem from "the failure to design human-machine interaction to exhibit the basic competencies of human-human interaction."[19]

Engineers and programmers compound the problems when they hide the workings of their creations from the operators, turning every system into an inscrutable black box. Normal human beings, the unstated assumption goes, don't have the smarts or the training to grasp the intricacies of a software program or robotic apparatus. If you tell them too much about the algorithms or procedures that govern its operations and decisions, you'll just confuse them or, worse yet, encourage them to tinker with the system. It's safer to keep people in the dark. Here again, though, the attempt to avoid human errors by removing personal responsibility ends up making the errors more likely. An ignorant operator is a dangerous operator. As the University of Iowa human-factors professor John Lee explains, it's common for an automated system to use "control algorithms that are at odds with

the control strategies and mental model of the person [operating it]."
If the person doesn't understand those algorithms, there's no way
she can "anticipate the actions and limits of the automation." The
human and the machine, operating under conflicting assumptions,
end up working at cross-purposes. People's inability to comprehend
the machines they use can also undermine their self-confidence,
Lee reports, which "can make them less inclined to intervene" when
something goes wrong.[20]

■   ■   ■   ■

HUMAN-FACTORS EXPERTS have long urged designers to move
away from the technology-first approach and instead embrace *human-centered automation*. Rather than beginning with an assessment of
the capabilities of the machine, human-centered design begins with
a careful evaluation of the strengths and limitations of the people
who will be operating or otherwise interacting with the machine. It
brings technological development back to the humanistic principles
that inspired the original ergonomists. The goal is to divide roles and
responsibilities in a way that not only capitalizes on the computer's
speed and precision but also keeps workers engaged, active, and
alert—in the loop rather than out of it.[21]

Striking that kind of balance isn't hard. Decades of ergonomic
research show it can be achieved in a number of straightforward
ways. A system's software can be programmed to shift control over
critical functions from the computer back to the operator at frequent
but irregular intervals. Knowing that they may need to take command at any moment keeps people attentive and engaged, promoting
situational awareness and learning. A design engineer can put limits
on the scope of automation, making sure that people working with
computers perform challenging tasks rather than being relegated to
passive, observational roles. Giving people more to do helps sustain

the generation effect. A designer can also give the operator direct sensory feedback on the system's performance, using audio and tactile alerts as well as visual displays, even for those activities that the computer is handling. Regular feedback heightens engagement and helps operators remain vigilant.

One of the most intriguing applications of the human-centered approach is *adaptive automation*. In adaptive systems, the computer is programmed to pay close attention to the person operating it. The division of labor between the software and the human operator is adjusted continually, depending on what's happening at any given moment.[22] When the computer senses that the operator has to perform a tricky maneuver, for example, it might take over all the other tasks. Freed from distractions, the operator can concentrate her full attention on the critical challenge. Under routine conditions, the computer might shift more tasks over to the operator, increasing her workload to ensure that she maintains her situational awareness and practices her skills. Putting the analytical capabilities of the computer to humanistic use, adaptive automation aims to keep the operator at the peak of the Yerkes-Dodson performance curve, preventing both cognitive overload and cognitive underload. DARPA, the Department of Defense laboratory that spearheaded the creation of the internet, is even working on developing "neuroergonomic" systems that, using various brain and body sensors, can "detect an individual's cognitive state and then manipulate task parameters to overcome perceptual, attentional, and working memory bottlenecks."[23] Adaptive automation also holds promise for injecting a dose of humanity into the working relationships between people and computers. Some early users of the systems report that they feel as though they're collaborating with a colleague rather than operating a machine.

Studies of automation have tended to focus on large, complex, and risk-laden systems, the kind used on flight decks, in control rooms, and on battlefields. When these systems fail, many lives and a great deal of money can be lost. But the research is also relevant to the design of decision-support applications used by doctors, lawyers, managers, and others in analytical trades. Such programs go through a lot of personal testing to make them easy to learn and operate, but once you dig beneath the user-friendly interface, you find that the technology-centered ethic still holds sway. "Typically," writes John Lee, "expert systems act as a prosthesis, supposedly replacing flawed and inconsistent human reasoning with more precise computer algorithms."[24] They're intended to supplant, rather than supplement, human judgment. With each upgrade in an application's data-crunching speed and predictive acumen, the programmer shifts more decision-making responsibility from the professional to the software.

Raja Parasuraman, who has studied the personal consequences of automation as deeply as anyone, believes this is the wrong approach. He argues that decision-support applications work best when they deliver pertinent information to professionals at the moment they need it, without recommending specific courses of action.[25] The smartest, most creative ideas come when people are afforded room to think. Lee agrees. "A less automated approach, which places the automation in the role of critiquing the operator, has met with much more success," he writes. The best expert systems present people with "alternative interpretations, hypotheses, or choices." The added and often unexpected information helps counteract the natural cognitive biases that sometimes skew human judgment. It pushes analysts and decision makers to look at problems from different perspectives and consider broader sets of options. But Lee stresses that the systems should leave the final verdict to the person. In the absence of perfect automation, he counsels, the evidence shows that "a lower level of automation, such as that used in the critiquing approach, is less likely to induce errors."[26] Computers do a superior job of sorting through lots of data quickly, but human experts remain subtler and wiser thinkers than their digital partners.

Carving out a protected space for the thoughts and judgments of expert practitioners is also a goal of those seeking a more humanistic approach to automation in the creative trades. Many designers criticize popular CAD programs for their pushiness. Ben Tranel, an architect with the Gensler firm in San Francisco, praises computers for expanding the possibilities of design. He points to the new, Gensler-designed Shanghai Tower in China, a spiraling, energy-efficient skyscraper, as an example of a building that "couldn't have been built" without computers. But he worries that the literalism of design software—the way it forces architects to define the meaning and use of every geometric element they input—is foreclosing the open-ended, unstructured explorations that freehand sketching encouraged. "A drawn line can be many things," he says, whereas a digitized line has to be just one thing.[27]

Back in 1996, the architecture professors Mark Gross and Ellen Yi-Luen Do proposed an alternative to literal-minded CAD software. They created a conceptual blueprint of an application with a "paper-like" interface that would be able to "capture users' intended ambiguity, vagueness, and imprecision and convey these qualities visually." It would lend design software "the suggestive power of the sketch."[28] Since then, many other scholars have made similar proposals. Recently, a team led by Yale computer scientist Julie Dorsey created a prototype of a design application that provides a "mental canvas." Rather than having the computer automatically translate two-dimensional drawings into three-dimensional virtual models, the system, which uses a touchscreen tablet as an input device, allows an architect to do rough sketches in three dimensions. "Designers can draw and redraw lines without being bound by the constraints of a polygonal mesh or the inflexibility of a parametric pipeline," the team explained. "Our system allows easy iterative refinement throughout the development of an idea, without imposing geometric precision before the idea is ready for it."[29] With less pushy software, a designer's imagination has more chance to flourish.

■   ■   ■   ■

THE TENSION between technology-centered and human-centered automation is not just a theoretical concern of academics. It affects decisions made every day by business executives, engineers and programmers, and government regulators. In the aviation business, the two dominant airliner manufacturers have been on different sides of the design question since the introduction of fly-by-wire systems thirty years ago. Airbus pursues a technology-centered approach. Its goal is to make its planes essentially "pilot-proof."[30] The company's decision to replace the bulky, front-mounted control yokes that have traditionally steered planes with diminutive, side-mounted joysticks was one expression of that goal. The game-like controllers send inputs to the flight computers efficiently, with minimal manual effort, but they don't provide pilots with tactile feedback. Consistent with the ideal of the glass cockpit, they emphasize the pilot's role as a computer operator rather than as an aviator. Airbus has also programmed its computers to override pilots' instructions in certain situations in order to keep the jet within the software-specified parameters of its flight envelope. The software, not the pilot, wields ultimate control.

Boeing has taken a more human-centered tack in designing its fly-by-wire craft. In a move that would have made the Wright brothers happy, the company decided that it wouldn't allow its flight software to override the pilot. The aviator retains final authority over maneuvers, even in extreme circumstances. And not only has Boeing kept the big yokes of yore; it has designed them to provide artificial feedback that mimics what pilots felt back when they had direct control over a plane's steering mechanisms. Although the yokes are just sending electronic signals to computers, they've been programmed to provide resistance and other tactile cues that simulate the feel of the movements of the plane's ailerons, elevators, and other control surfaces. Research has found that tactile, or haptic, feedback is significantly more effective than visual cues alone in alerting pilots to important changes in a plane's orientation and operation, according to John Lee. And because the brain processes tactile signals in a different way than visual signals, "haptic warnings" don't tend to "interfere with the performance of concurrent visual tasks."[31] In a sense, the synthetic, tactile feedback takes Boeing pilots out of the glass cockpit. They may not wear their jumbo jets the way Wiley Post wore his little Lockheed Vega, but they are more involved in the bodily experience of flight than are their counterparts on Airbus flight decks.

Airbus makes magnificent planes. Some commercial pilots prefer them to Boeing's jets, and the safety records of the two manufacturers are pretty much identical. But recent incidents reveal the shortcomings of Airbus's technology-centered approach. Some aviation experts believe that the design of the Airbus cockpit played a part in the Air France disaster. The voice-recorder transcript revealed that the whole time the pilot controlling the plane, Pierre-Cédric Bonin, was pulling back on his sidestick, his copilot, David Robert, was oblivious to Bonin's fateful mistake. In a Boeing cockpit, each pilot has a clear view of the other pilot's yoke and how it's being handled. If that weren't enough, the two yokes operate as a single unit. If one pilot pulls back on his yoke, the other pilot's goes back too. Through both visual and haptic cues, the pilots stay in sync. The Airbus sidesticks, in contrast, are not in clear view, they work with much subtler motions, and they operate independently. It's easy for a pilot to miss what his colleague is doing, particularly in emergencies when stress rises and focus narrows.

Had Robert seen and corrected Bonin's error early on, the pilots may well have regained control of the A330. The Air France crash, Chesley Sullenberger has said, would have been "much less likely to happen" if the pilots had been flying in a Boeing cockpit with its human-centered controls.[32] Even Bernard Ziegler, the brilliant and proud French engineer who served as Airbus's top designer until his retirement in 1997, recently expressed misgivings about his company's design philosophy. "Sometimes I wonder if we made an airplane that is too easy to fly," he said to William Langewiesche, the writer, during an interview in Toulouse, where Airbus has its headquarters. "Because in a difficult airplane the crews may stay more alert." He went on to suggest that Airbus "should have built a kicker into the pilots' seats."[33] He may have been joking, but his comment jibes with what human-factors researchers have learned about the maintenance of human skills and attentiveness. Sometimes a good kick, or its technological equivalent, is exactly what an automated system needs to give its operators.

When the FAA, in its 2013 safety alert for operators, suggested that airlines encourage pilots to assume manual control of their planes more frequently during flights, it was also taking a stand, if a tentative one, in favor of human-centered automation. Keeping the pilot more firmly in the loop, the agency had come to realize, could reduce the chances of human error, temper the consequences of automation failure, and make air travel even safer than it already is. More automation is not always the wisest choice. The FAA, which employs a large and respected group of human-factors researchers, is also paying close attention to ergonomics as it plans its ambitious "NextGen" overhaul of the nation's air-traffic-control system. One of the project's overarching goals is to "create aerospace systems that adapt to, compensate for, and augment the performance of the human."[34]

In the financial industry, the Royal Bank of Canada is also going against the grain of technology-centered automation. At its Wall Street trading desk, it has installed a proprietary software program, called THOR, that actually slows down the transmission of buy and sell orders in a way that protects them from the algorithmic manipulations of high-speed traders. By slowing the orders, RBC has found, trades often end up being executed at more attractive terms for its customers. The bank admits that it's making a trade-off in resisting the prevailing technological imperative of speedy data flows. By eschewing high-speed trading, it makes a little less money on each trade. But it believes that, over the long run, the strengthening of client loyalty and the reduction of risk will lead to higher profits overall.[35]

One former RBC executive, Brad Katsuyama, is going even further. Having watched stock markets become skewed in favor of high-frequency traders, he spearheaded the creation of a new and fairer exchange, called IEX. Opened late in 2013, IEX imposes controls on automated systems. Its software manages the flow of data to ensure that all members of the exchange receive pricing and other information at the same time, neutralizing the advantages enjoyed by predatory trading firms that situate their computers next door to exchanges. And IEX forbids certain kinds of trades and fee schemes that give an edge to speedy algorithms. Katsuyama and his colleagues are using sophisticated technology to level the playing field between people and computers. Some national regulatory agencies are also trying to put the brakes on automated trading, through laws and regulations. In 2012, France placed a small tax on stock trades, and Italy followed suit a year later. Because high-frequency-trading algorithms are usually designed to execute volume-based arbitrage strategies—each trade returns only a minuscule profit, but millions of trades are made in a matter of moments—even a tiny transaction tax can render the programs much less attractive.

■   ■   ■   ■

SUCH ATTEMPTS to rein in automation are encouraging. They show that at least some businesses and government agencies are willing to question the prevailing technology-first attitude. But these efforts remain exceptions to the rule, and their continued success is far from assured. Once technology-centered automation has taken hold in a field, it becomes very hard to alter the course of progress. The software comes to shape how work is done, how operations are organized, what consumers expect, and how profits are made. It becomes an economic and a social fixture. This process is an example of what the historian Thomas Hughes calls "technological momentum."[36] In its early development, a new technology is malleable; its form and use can be shaped not only by the desires of its designers but also by the concerns of those who use it and the interests of society as a whole. But once the technology becomes embedded in physical infrastructure, commercial and economic arrangements, and personal and political norms and expectations, changing it becomes enormously difficult. The technology is at that point an integral component of the social status quo. Having amassed great inertial force, it continues down the path it's on. Particular technological components will still become outdated, of course, but they'll tend to be replaced by new ones that refine and perpetuate the existing modes of operation and the related measures of performance and success.

The commercial aviation system, for example, now depends on the precision of computer control. Computers are better than pilots at plotting the most fuel-efficient routes, and computer-controlled planes can fly closer together than can planes operated by people. There's a fundamental tension between the desire to enhance pilots' manual flying skills and the pursuit of ever higher levels of automation in the skies. Airlines are unlikely to sacrifice profits and regulators are unlikely to curtail the capacity of the aviation system in order to give pilots significantly more time to practice flying by hand. The rare automation-related disaster, however horrifying, may be accepted as a cost of an efficient and profitable transport system. In health care, insurers and hospital companies, not to mention politicians, look to automation as a quick fix to lower costs and boost productivity. They'll almost certainly keep ratcheting up the pressure on providers to automate medical practices and procedures in order to save money, even if doctors have worries about the long-term erosion of their most subtle and valuable talents. On financial exchanges, computers can execute a trade in ten milliseconds—that's one hundredth of a second—but it takes the human brain nearly a quarter of a second to respond to an event or other stimulus. A computer can process tens of thousands of trades in the blink of a trader's eye.[37] The speed of the computer has taken the person out of the picture.

It's commonly assumed that any technology that comes to be broadly adopted in a field, and hence gains momentum, must be the best one for the job. Progress, in this view, is a quasi-Darwinian process. Many different technologies are invented, they compete for users and buyers, and after a period of rigorous testing and comparison the marketplace chooses the best of the bunch. Only the fittest tools survive. Society can thus be confident that the technologies it employs are the optimum ones—and that the alternatives discarded along the way were flawed in some fatal way. It's a reassuring view of progress, founded on, in the words of the late historian David Noble, "a simple faith in objective science, economic rationality, and the market." But as Noble went on to explain in his 1984 book *Forces of Production*, it's a distorted view: "It portrays technological development as an autonomous and neutral technical process, on the one hand, and a coldly rational and self-regulating process, on the other, neither of which accounts for people, power, institutions, competing values, or different dreams."[38] In place of the complexities, vagaries, and intrigues of history, the prevailing view of technological progress presents us with a simplistic, retrospective fantasy.

Noble illustrated the tangled way technologies actually gain acceptance and momentum through the story of the automation of the machine tool industry in the years after World War II. Inventors and engineers developed several different techniques for programming lathes, drill presses, and other factory tools, and each of the control methods had advantages and disadvantages. One of the simplest and most ingenious of the systems, called Specialmatic, was invented by a Princeton-trained engineer named Felix P. Caruthers and marketed by a small New York company called Automation Specialties. Using an array of keys and dials to encode and control the workings of a machine, Specialmatic put the power of programming into the hands of skilled machinists on the factory floor. A machine operator, explained Noble, "could set and adjust feeds and speeds, relying upon accumulated experience with the sights, sounds, and smells of metal cutting."[39] In addition to bringing the tacit know-how of the experienced craftsman into the automated system, Specialmatic had an economic advantage: a manufacturer did not have to pay a squad of engineers and consultants to program its equipment. Caruthers's technology earned accolades from *American Machinist* magazine, which noted that Specialmatic "is designed to permit complete set-up and programming at the machine." It would allow the machinist to gain the efficiency benefits of automation while retaining "full control of his machine throughout its entire machining cycle."[40]

But Specialmatic never gained a foothold in the market. While Caruthers was working on his invention, the U.S. Air Force was plowing money into a research program, conducted by an MIT team with long-standing ties to the military, to develop "numerical control," a digital coding technique that was a forerunner of modern software programming. Not only did numerical control enjoy the benefits of a generous government subsidy and a prestigious academic pedigree; it appealed to business owners and managers who, faced with unremitting labor tensions, yearned to gain more control over the operation of machinery in order to undercut the power of workers and their unions. Numerical control also had the glow of a cutting-edge technology—it was carried along by the burgeoning postwar excitement over digital computers. The MIT system may have been, as the author of a Society of Manufacturing Engineers paper would later write, "a complicated, expensive monstrosity,"[41] but industrial giants like GE and Westinghouse rushed to embrace the technology, never giving alternatives like Specialmatic a chance. Far from winning a tough evolutionary battle for survival, numerical control was declared the victor before competition even began. Programming took precedence over people, and the momentum behind the technology-first design philosophy grew. As for the general public, it never knew that a choice had been made.

Engineers and programmers shouldn't bear all the blame for the ill effects of technology-centered automation. They may be guilty at times of pursuing narrowly mechanistic dreams and desires, and they may be susceptible to the "technical arrogance" that "gives people an illusion of illimitable power," in the words of the physicist Freeman Dyson.[42] But they're also responding to the demands of employers and clients. Software developers always face a trade-off in writing programs for automating work. Taking the steps necessary to promote the development of expertise—restricting the scope of automation, giving a greater and more active role to people, encouraging the development of automaticity through rehearsal and repetition—entails a sacrifice of speed and yield. Learning requires inefficiency. Businesses, which seek to maximize productivity and profit, would rarely, if ever, accept such a trade-off. The main reason they invest in automation, after all, is to reduce labor costs and streamline operations.

As individuals, too, we almost always seek efficiency and convenience when we decide which software application or computing device to use. We pick the program or gadget that lightens our load and frees up our time, not the one that makes us work harder and longer. Technology companies naturally cater to such desires when they design their wares. They compete fiercely to offer the product that requires the least effort and thought to use. "At Google and all these places," says Google executive Alan Eagle, explaining the guiding philosophy of many software and internet businesses, "we make technology as brain-dead easy to use as possible."[43] When it comes to the development and use of commercial software, whether it underpins an industrial system or a smartphone app, abstract concerns about the fate of human talent can't compete with the prospect of saving time and money.

I asked Parasuraman whether he thinks society will come to use automation more wisely in the future, striking a better balance between computer calculation and personal judgment, between the pursuit of efficiency and the development of expertise. He paused a moment and then, with a wry laugh, said, "I'm not very sanguine."

Interlude,
with Grave Robber

1

I WAS IN A FIX. I HAD—BY NECESSITY, NOT CHOICE—STRUCK
up an alliance with a demented grave robber named Seth Briars. "I
don't eat, I don't sleep, I don't wash, and I don't care," Seth had informed me, not without a measure of pride, shortly after we met in
the cemetery beside Coot's Chapel. He knew the whereabouts of
certain individuals I was seeking, and in exchange for leading me to
them, he had demanded that I help him cart a load of fresh corpses
out past Critchley's Ranch to a dusty ghost town called Tumbleweed.
I drove Seth's horse-drawn wagon, while he stayed in the back, ri-
fling the dead for valuables. The trip was a trial. We made it through
an ambush by highwaymen along the route—with firearms, I was
more than handy—but when I tried to cross a rickety bridge near
Gaptooth Ridge, the weight of the bodies shifted and I lost control
of the horses. The wagon careened into a ravine, and I died in a
volcanic, screen-coating eruption of blood. I came back to life after
a couple of purgatorial seconds, only to go through the ordeal again.
After a half-dozen failed attempts, I began to despair of ever completing the mission.

The game I was playing, an exquisitely crafted, goofily written open-world shooter called Red Dead Redemption, is set in the early years of the last century, in a mythical southwestern border territory named New Austin. Its plot is pure Peckinpah. When you start the game, you assume the role of a stoic outlaw-turned-rancher named John Marston, whose right cheek is riven by a couple of long, symbolically deep scars. Marston is being blackmailed into tracking down his old criminal associates by federal agents who are holding his wife and young son hostage. To complete the game, you have to guide the gunslinger through various feats of skill and cunning, each a little tougher than the one preceding it.

After a few more tries, I finally did make it over that bridge, grisly cargo in tow. In fact, after many mayhem-filled hours in front of my Xbox-connected flat-screen TV, I managed to get through all of the game's fifty-odd missions. As my reward, I got to watch myself—John Marston, that is—be gunned down by the very agents who had forced him into the quest. Gruesome ending aside, I came away from the game with a feeling of accomplishment. I had roped mustangs, shot and skinned coyotes, robbed trains, won a small fortune playing poker, fought alongside Mexican revolutionaries, rescued harlots from drunken louts, and, in true *Wild Bunch* fashion, used a Gatling gun to send an army of thugs to Kingdom Come. I had been tested, and my middle-aged reflexes had risen to the challenge. It may not have been an epic win, but it was a win.

Video games tend to be loathed by people who have never played them. That's understandable, given the gore involved, but it's a shame. In addition to their considerable ingenuity and occasional beauty, the best games provide a model for the design of software. They show how applications can encourage the development of skills rather than their atrophy. To master a video game, a player has to struggle through challenges of increasing difficulty, always pushing the limits of his talent. Every mission has a goal, there are rewards for doing well, and the feedback (an eruption of blood, perhaps) is immediate and often visceral. Games promote a state of flow, inspiring players to repeat tricky maneuvers until they become second nature. The skill a gamer learns may be trivial—how to manipulate a plastic controller to drive an imaginary wagon over an imaginary bridge, say—but he'll learn it thoroughly, and he'll be able to exercise it again in the next mission or the next game. He'll become an expert, and he'll have a blast along the way.*

When it comes to the software we use in our personal lives, video games are an exception. Most popular apps, gadgets, and online services are built for convenience, or, as their makers say, "usability." Requiring only a few taps, swipes, or clicks, the programs can be mastered with little study or practice. Like the automated systems used in industry and commerce, they've been carefully designed to shift the burden of thought from people to computers. Even the high-end programs used by musicians, record producers, filmmakers, and photographers place an ever stronger emphasis on ease of use. Complex audio and visual effects, which once demanded expert know-how, can be achieved by pushing a button or dragging a slider.

---

* In suggesting video games as a model for programmers, I'm not endorsing the voguish software-design practice that goes by the ugly name "gamification." That's when an app or a website uses a game-like reward system to motivate or manipulate people into repeating some prescribed activity. Building on the operant-conditioning experiments of the psychologist B. F. Skinner, gamification exploits the flow state's dark side. Seeking to sustain the pleasures and rewards of flow, people can become obsessive in their use of the software. Computerized slot machines, to take one notorious example, are carefully designed to promote an addictive form of flow in their players, as Natasha Dow Schüll describes in her chilling book *Addiction by Design: Machine Gambling in Vegas* (Princeton: Princeton University Press, 2012). An experience that is normally "life affirming, restorative, and enriching," she writes, becomes for gamblers "depleting, entrapping, and associated with a loss of autonomy." Even when used for ostensibly benign purposes, such as dieting, gamification wields a cynical power. Far from being an antidote to technology-centered design, it takes the practice to an extreme. It seeks to automate human will.

The underlying concepts need not be understood, as they've been incorporated into software routines. This has the very real benefit of making the software useful to a broader group of people—those who want to get the effects without the effort. But the cost of accommodating the dilettante is a demeaning of expertise.

Peter Merholz, a respected software-design consultant, counsels programmers to seek "frictionlessness" and "simplicity" in their products. Successful devices and applications, he says, hide their technical complexity behind user-friendly interfaces. They minimize the cognitive load they place on users: "Simple things don't require a lot of thought. Choices are eliminated, recall is not required."[1] That's a recipe for creating the kinds of applications that, as Christof van Nimwegen's Cannibals and Missionaries experiment demonstrated, bypass the mental processes of learning, skill building, and memorization. The tools demand little of us and, cognitively speaking, give little to us.

What Merholz calls the "it just works" design philosophy has a lot going for it. Anyone who has struggled to set the alarm on a digital clock or change the settings on a WiFi router or figure out Microsoft Word's toolbars knows the value of simplicity. Needlessly complicated products waste time without much compensation. It's true we don't need to be experts at everything, but as software writers take to scripting processes of intellectual inquiry and social attachment, frictionlessness becomes a problematic ideal. It can sap us not only of know-how but of our sense that know-how is something important and worth cultivating. Think of the algorithms for reviewing and correcting spelling that are built into virtually every writing and messaging application these days. Spell checkers once served as tutors. They'd highlight possible errors, calling your attention to them and, in the process, giving you a little spelling lesson. You learned as you used them. Now, the tools incorporate autocorrect functions. They instantly and surreptitiously clean up your mistakes, without alerting you to them. There's no feedback, no "friction." You see nothing and learn nothing.

Or think of Google's search engine. In its original form, it presented you with nothing but an empty text box. The interface was a model of simplicity, but the service still required you to think about your query, to consciously compose and refine a set of keywords to get the best results. That's no longer necessary. In 2008, the company introduced Google Suggest, an autocomplete routine that uses prediction algorithms to anticipate what you're looking for. Now, as soon as you type a letter into the search box, Google offers a set of suggestions for how to phrase your query. With each succeeding letter, a new set of suggestions pops up. Underlying the company's hyperactive solicitude is a dogged, almost monomaniacal pursuit of efficiency. Taking the misanthropic view of automation, Google has come to see human cognition as creaky and inexact, a cumbersome biological process better handled by a computer. "I envision some years from now that the majority of search queries will be answered without you actually asking," says Ray Kurzweil, the inventor and futurist who in 2012 was appointed Google's director of engineering. The company will "just know this is something that you're going to want to see."[2] The ultimate goal is to fully automate the act of searching, to take human volition out of the picture.

Social networks like Facebook seem impelled by a similar aspiration. Through the statistical "discovery" of potential friends, the provision of "Like" buttons and other clickable tokens of affection, and the automated management of many of the time-consuming aspects of personal relations, they seek to streamline the messy process of affiliation. Facebook's founder, Mark Zuckerberg, celebrates all of this as "frictionless sharing"—the removal of conscious effort from socializing. But there's something repugnant about applying the bureaucratic ideals of speed, productivity, and standardization to our relations with others. The most meaningful bonds aren't forged through transactions in a marketplace or other routinized exchanges of data. People aren't nodes on a network grid. The bonds require trust and courtesy and sacrifice, all of which, at least to a technocrat's mind, are sources of inefficiency and inconvenience. Removing the friction from social attachments doesn't strengthen them; it weakens them. It makes them more like the attachments between consumers and products—easily formed and just as easily broken.

Like meddlesome parents who never let their kids do anything on their own, Google, Facebook, and other makers of personal software end up demeaning and diminishing qualities of character that, at least in the past, have been seen as essential to a full and vigorous life: ingenuity, curiosity, independence, perseverance, daring. It may be that in the future we'll only experience such virtues vicariously, through the exploits of action figures like John Marston in the fantasy worlds we enter through screens.

CHAPTER EIGHT

YOUR INNER DRONE

IT's A COLD, MISTY FRIDAY NIGHT IN MID-DECEMBER AND you're driving home from your office holiday party. Actually, you're being driven home. You recently bought your first autonomous car—a Google-programmed, Mercedes-built eSmart electric sedan—and the software is at the wheel. You can see from the glare of your self-adjusting LED headlights that the street is icy in spots, and you know, thanks to the continuously updated dashboard display, that the car is adjusting its speed and traction settings accordingly. All's going smoothly. You relax and let your mind drift back to the evening's stilted festivities. But as you pass through a densely wooded stretch of road, just a few hundred yards from your driveway, an animal darts into the street and freezes, directly in the path of the car. It's your neighbor's beagle, you realize—the one that's always getting loose.

What does your robot driver do? Does it slam on the brakes, in hopes of saving the dog but at the risk of sending the car into an uncontrolled skid? Or does it keep its virtual foot off the brake, sacrificing the beagle to ensure that you and your vehicle stay out of harm's way? How does it sort through and weigh the variables and probabilities to arrive at a split-second decision? If its algorithms calculate that hitting the brakes would give the dog a 53 percent chance of survival but would entail an 18 percent chance of damaging the car and a 4 percent chance of causing injury to you, does it conclude that trying to save the animal would be the right thing to do? How does the software, working on its own, translate a set of numbers into a decision that has both practical and moral consequences?

What if the animal in the road isn't your neighbor's pet but your own? What, for that matter, if it isn't a dog but a child? Imagine you're on your morning commute, scrolling through your overnight emails as your self-driving car crosses a bridge, its speed precisely synced to the forty-mile-per-hour limit. A group of schoolchildren is also heading over the bridge, on the pedestrian walkway that runs alongside your lane. The kids, watched by adults, seem orderly and well behaved. There's no sign of trouble, but your car slows slightly, its computer preferring to err on the side of safety. Suddenly, there's a tussle, and a little boy is pushed into the road. Busily tapping out a message on your smartphone, you're oblivious to what's happening. Your car has to make the decision: either it swerves out of its lane and goes off the opposite side of the bridge, possibly killing you, or it hits the child. What does the software instruct the steering wheel to do? Would the program make a different choice if it knew that one of your own children was riding with you, strapped into a sensor-equipped car seat in the back? What if there was an oncoming vehicle in the other lane? What if that vehicle was a school bus? Isaac Asimov's first law of robot ethics—"a robot may not injure a human being, or, through inaction, allow a human being to come to harm"[1]—sounds reasonable and reassuring, but it assumes a world far simpler than our own.

The arrival of autonomous vehicles, says Gary Marcus, the NYU psychology professor, would do more than "signal the end of one more human niche." It would mark the start of a new era in which machines will have to have "ethical systems."[2] Some would argue that we're already there. In small but ominous ways, we have started handing off moral decisions to computers. Consider Roomba, the much-publicized robotic vacuum cleaner. Roomba makes no distinction between a dust bunny and an insect. It gobbles both, indiscriminately. If a cricket crosses its path, the cricket gets sucked to its death. A lot of people, when vacuuming, will also run over the cricket. They place no value on a bug's life, at least not when the bug is an intruder in their home. But other people will stop what they're doing, pick up the cricket, carry it to the door, and set it loose. (Followers of Jainism, the ancient Indian religion, consider it a sin to harm any living thing; they take great care not to kill or hurt insects.) When we set Roomba loose on a carpet, we cede to it the power to make moral choices on our behalf. Robotic lawn mowers, like Lawn-Bott and Automower, routinely deal death to higher forms of life, including reptiles, amphibians, and small mammals. Most people, when they see a toad or a field mouse ahead of them as they cut their grass, will make a conscious decision to spare the animal, and if they should run it over by accident, they'll feel bad about it. A robotic lawn mower kills without compunction.

Up to now, discussions about the morals of robots and other machines have been largely theoretical, the stuff of science-fiction stories or thought experiments in philosophy classes. Ethical considerations have often influenced the design of tools—guns have safeties, motors have governors, search engines have filters—but machines haven't been required to have consciences. They haven't had to adjust their own operation in real time to account for the ethical vagaries of a situation. Whenever questions about the moral use of a technology arose in the past, people would step in to sort things out. That won't always be feasible in the future. As robots and computers become more adept at sensing the world and acting autonomously in it, they'll inevitably face situations in which there's no one right choice. They'll have to make vexing decisions on their own. It's impossible to automate complex human activities without also automating moral choices.

Human beings are anything but flawless when it comes to ethical judgments. We frequently do the wrong thing, sometimes out of confusion or heedlessness, sometimes deliberately. That's led some to argue that the speed with which robots can sort through options, estimate probabilities, and weigh consequences will allow them to make more rational choices than people are capable of making when immediate action is called for. There's truth in that view. In certain circumstances, particularly those where only money or property is at stake, a swift calculation of probabilities may be sufficient to determine the action that will lead to the optimal outcome. Some human drivers will try to speed through a traffic light that's just turning red, even though it ups the odds of an accident. A computer would never act so rashly. But most moral dilemmas aren't so tractable. Try to solve them mathematically, and you arrive at a more fundamental question: Who determines what the "optimal" or "rational" choice is in a morally ambiguous situation? Who gets to program the robot's conscience? Is it the robot's manufacturer? The robot's owner? The software coders? Politicians? Government regulators? Philosophers? An insurance underwriter?

There is no perfect moral algorithm, no way to reduce ethics to a set of rules that everyone will agree on. Philosophers have tried to do that for centuries, and they've failed. Even coldly utilitarian calculations are subjective; their outcome hinges on the values and interests of the decision maker. The rational choice for your car's insurer—the dog dies—might not be the choice you'd make, either deliberately or reflexively, when you're about to run over a neighbor's pet. "In an age of robots," observes the political scientist Charles Rubin, "we will be as ever before—or perhaps as never before—stuck with morality."[3]

Still, the algorithms will need to be written. The idea that we can calculate our way out of moral dilemmas may be simplistic, or even repellent, but that doesn't change the fact that robots and software agents are going to have to calculate their way out of moral dilemmas. Unless and until artificial intelligence attains some semblance of consciousness and is able to feel or at least simulate emotions like affection and regret, no other course will be open to our calculating kin. We may rue the fact that we've succeeded in giving automatons the ability to take moral action before we've figured out how to give them moral sense, but regret doesn't let us off the hook. The age of ethical systems is upon us. If autonomous machines are to be set loose in the world, moral codes will have to be translated, however imperfectly, into software codes.

■   ■   ■   ■

HERE'S ANOTHER scenario. You're an army colonel who's commanding a battalion of human and mechanical soldiers. You have a platoon of computer-controlled "sniper robots" stationed on street corners and rooftops throughout a city that your forces are defending against a guerrilla attack. One of the robots spots, with its laser-vision sight, a man in civilian clothes holding a cell phone. He's acting in a way that experience would suggest is suspicious. The robot, drawing on a thorough analysis of the immediate situation and a rich database documenting past patterns of behavior, instantly calculates that there's a 68 percent chance the person is an insurgent preparing to detonate a bomb and a 32 percent chance he's an innocent bystander. At that moment, a personnel carrier is rolling down the street with a dozen of your human soldiers on board. If there is a bomb, it could be detonated at any moment. War has no pause button. Human judgment can't be brought to bear. The robot has to act. What does its software order its gun to do: shoot or hold fire?

If we, as civilians, have yet to grapple with the ethical implications of self-driving cars and other autonomous robots, the situation is very different in the military. For years, defense departments and military academies have been studying the methods and consequences of handing authority for life-and-death decisions to battlefield machines. Missile and bomb strikes by unmanned drone aircraft, such as the Predator and the Reaper, are already commonplace, and they've been the subject of heated debates. Both sides make good arguments. Proponents note that drones keep soldiers and airmen out of harm's way and, through the precision of their attacks, reduce the casualties and damage that accompany traditional combat and bombardment. Opponents see the strikes as state-sponsored assassinations. They point out that the explosions frequently kill or wound, not to mention terrify, civilians. Drone strikes, though, aren't automated; they're remote-controlled. The planes may fly themselves and perform surveillance functions on their own, but decisions to fire their weapons are made by soldiers sitting at computers and monitoring live video feeds, operating under strict orders from their superiors. As currently deployed, missile-carrying drones aren't all that different from cruise missiles and other weapons. A person still pulls the trigger.

The big change will come when a computer starts pulling the trigger. Fully automated, computer-controlled killing machines—what the military calls lethal autonomous robots, or LARs—are technologically feasible today, and have been for quite some time. Environmental sensors can scan a battlefield with high-definition precision, automatic firing mechanisms are in wide use, and codes to control the shooting of a gun or the launch of a missile aren't hard to write. To a computer, a decision to fire a weapon isn't really any different from a decision to trade a stock or direct an email message into a spam folder. An algorithm is an algorithm.

In 2013, Christof Heyns, a South African legal scholar who serves as special rapporteur on extrajudicial, summary, and arbitrary executions to the United Nations General Assembly, issued a report on the status of and prospects for military robots.[4] Clinical and measured, it made for chilling reading. "Governments with the ability to produce LARs," Heyns wrote, "indicate that their use during armed conflict or elsewhere is not currently envisioned." But the history of weaponry, he went on, suggests we shouldn't put much stock in these assurances: "It should be recalled that aeroplanes and drones were first used in armed conflict for surveillance purposes only, and offensive use was ruled out because of the anticipated adverse consequences. Subsequent experience shows that when technology that provides a perceived advantage over an adversary is available, initial intentions are often cast aside." Once a new type of weaponry is deployed, moreover, an arms race almost always ensues. At that point, "the power of vested interests may preclude efforts at appropriate control."

War is in many ways more cut-and-dried than civilian life. There are rules of engagement, chains of command, well-demarcated sides. Killing is not only acceptable but encouraged. Yet even in war the programming of morality raises problems that have no solution—or at least can't be solved without setting a lot of moral considerations aside. In 2008, the U.S. Navy commissioned the Ethics and Emerging Sciences Group at California Polytechnic State University to prepare a white paper reviewing the ethical issues raised by LARs and laying out possible approaches to "constructing ethical autonomous robots" for military use. The ethicists reported that there are two basic ways to program a robot's computer to make moral decisions: top-down and bottom-up. In the top-down approach, all the rules governing the robot's decisions are programmed ahead of time, and the robot simply obeys the rules "without change or flexibility." That sounds straightforward, but it's not, as Asimov discovered when he tried to formulate his system of robot ethics. There's no way to anticipate all the circumstances a robot may encounter. The "rigidity" of top-down programming can backfire, the scholars wrote, "when events and situations unforeseen or insufficiently imagined by the programmers occur, causing the robot to perform badly or simply do horrible things, precisely because it is rule-bound."[5]

In the bottom-up approach, the robot is programmed with a few rudimentary rules and then sent out into the world. It uses machine-learning techniques to develop its own moral code, adapting it to new situations as they arise. "Like a child, a robot is placed into variegated situations and is expected to learn through trial and error (and feedback) what is and is not appropriate to do." The more dilemmas it faces, the more fine-tuned its moral judgment becomes. But the bottom-up approach presents even thornier problems. First, it's impracticable; we have yet to invent machine-learning algorithms subtle and robust enough for moral decision making. Second, there's no room for trial and error in life-and-death situations; the approach itself would be immoral. Third, there's no guarantee that the morality a computer develops would reflect or be in harmony with human morality. Set loose on a battlefield with a machine gun and a set of machine-learning algorithms, a robot might go rogue.

Human beings, the ethicists pointed out, employ a "hybrid" of top-down and bottom-up approaches in making moral decisions. People live in societies that have laws and other strictures to guide and control behavior; many people also shape their decisions and actions to fit religious and cultural precepts; and personal conscience, whether innate or not, imposes its own rules. Experience plays a role too. People learn to be moral creatures as they grow up and struggle with ethical decisions of different stripes in different situations. We're far from perfect, but most of us have a discriminating moral sense that can be applied flexibly to dilemmas we've never encountered before. The only way for robots to become truly moral beings would be to follow our example and take a hybrid approach, both obeying rules and learning from experience. But creating a machine with that capacity is far beyond our technological grasp. "Eventually," the ethicists con-

cluded, "we may be able to build morally intelligent robots that maintain the dynamic and flexible morality of bottom-up systems capable
of accommodating diverse inputs, while subjecting the evaluation of
choices and actions to top-down principles." Before that happens,
though, we'll need to figure out how to program computers to display
"supra-rational faculties"—to have emotions, social skills, consciousness, and a sense of "being embodied in the world."[6] We'll need to
become gods, in other words.

Armies are unlikely to wait that long. In an article in *Parameters*,
the journal of the U.S. Army War College, Thomas Adams, a military strategist and retired lieutenant colonel, argues that "the logic
leading to fully autonomous systems seems inescapable." Thanks to
the speed, size, and sensitivity of robotic weaponry, warfare is "leaving the realm of human senses" and "crossing outside the limits of
human reaction times." It will soon be "too complex for real human
comprehension." As people become the weakest link in the military
system, he says, echoing the technology-centric arguments of civilian
software designers, maintaining "meaningful human control" over
battlefield decisions will become next to impossible. "One answer,
of course, is to simply accept a slower information-processing rate as
the price of keeping humans in the military decision business. The
problem is that some adversary will inevitably decide that the way to
defeat the human-centric systems is to attack it with systems that are
not so limited." In the end, Adams believes, we "may come to regard
tactical warfare as properly the business of machines and not appropriate for people at all."[7]

What will make it especially difficult to prevent the deployment
of LARs is not just their tactical effectiveness. It's also that their
deployment would have certain ethical advantages independent of
the machines' own moral makeup. Unlike human fighters, robots
have no baser instincts to tug at them in the heat and chaos of battle.
They don't experience stress or depression or surges of adrenaline.

"Typically," Christof Heyns wrote, "they would not act out of revenge, panic, anger, spite, prejudice or fear. Moreover, unless specifically programmed to do so, robots would not cause intentional suffering on civilian populations, for example through torture. Robots also do not rape."[8]

Robots don't lie or otherwise try to hide their actions, either. They can be programmed to leave digital trails, which would tend to make an army more accountable for its actions. Most important of all, by using LARs to wage war, a country can avoid death or injury to its own soldiers. Killer robots save lives as well as take them. As soon as it becomes clear to people that automated soldiers and weaponry will lower the likelihood of their sons and daughters being killed or maimed in battle, the pressure on governments to automate war making may become irresistible. That robots lack "human judgement, common sense, appreciation of the larger picture, understanding of the intentions behind people's actions, and understanding of values," in Heyns's words, may not matter in the end. In fact, the moral stupidity of robots has its advantages. If the machines displayed human qualities of thought and feeling, we'd be less sanguine about sending them to their destruction in war.

The slope gets only more slippery. The military and political advantages of robot soldiers bring moral quandaries of their own. The deployment of LARs won't just change the way battles and skirmishes are fought, Heyns pointed out. It will change the calculations that politicians and generals make about whether to go to war in the first place. The public's distaste for casualties has always been a deterrent to fighting and a spur to negotiation. Because LARs will reduce the "human costs of armed conflict," the public may "become increasingly disengaged" from military debates and "leave the decision to use force as a largely financial or diplomatic question for the State, leading to the 'normalization' of armed conflict. LARs may thus lower the threshold for States for going to war or otherwise using lethal force, resulting in armed conflict no longer being a measure of last resort."[9]

The introduction of a new class of armaments always alters the nature of warfare, and weapons that can be launched or detonated from afar—catapults, mines, mortars, missiles—tend to have the greatest effects, both intended and unintended. The consequences of autonomous killing machines would likely go beyond anything that's come before. The first shot freely taken by a robot will be a shot heard round the world. It will change war, and maybe society, forever.

■   ■   ■   ■

THE SOCIAL and ethical challenges posed by killer robots and self-driving cars point to something important and unsettling about where automation is headed. The substitution myth has traditionally been defined as the erroneous assumption that a job can be divided into separate tasks and those tasks can be automated piecemeal without changing the nature of the job as a whole. That definition may need to be broadened. As the scope of automation expands, we're learning that it's also a mistake to assume that society can be divided into discrete spheres of activity—occupations or pastimes, say, or domains of governmental purview—and those spheres can be automated individually without changing the nature of society as a whole. Everything is connected—change the weapon, and you change the war—and the connections tighten when they're made explicit in computer networks. At some point, automation reaches a critical mass. It begins to shape society's norms, assumptions, and ethics. People see themselves and their relations to others in a different light, and they adjust their sense of personal agency and responsibility to account for technology's expanding role. They behave differently too. They *expect* the aid of computers, and on those rare instances when it's not forthcoming, they feel bewildered. Software takes on what the MIT computer scientist Joseph Weizenbaum termed a "compelling urgency." It becomes "the very stuff out of which man builds his world."[10]

In the 1990s, just as the dot-com bubble was beginning to inflate, there was much excited talk about "ubiquitous computing." Soon, pundits assured us, microchips would be everywhere—embedded in factory machinery and warehouse shelving, affixed to the walls of offices and shops and homes, buried in the ground and floating in the air, installed in consumer goods and woven into clothing, even swimming around in our bodies. Equipped with sensors and transceivers, the tiny computers would measure every variable imaginable, from metal fatigue to soil temperature to blood sugar, and they'd send their readings, via the internet, to data-processing centers, where bigger computers would crunch the numbers and output instructions for keeping everything in spec and in sync. Computing would be pervasive; automation, ambient. We'd live in a geek's paradise, the world a programmable machine.

One of the main sources of the hype was Xerox PARC, the fabled Silicon Valley research lab where Steve Jobs found the inspiration for the Macintosh. PARC's engineers and information scientists published a series of papers portraying a future in which computers would be so deeply woven into "the fabric of everyday life" that they'd be "indistinguishable from it."[11] We would no longer even notice all the computations going on around us. We'd be so saturated with data, so catered to by software, that, instead of experiencing the anxiety of information overload, we'd feel "encalmed."[12] It sounded idyllic. But the PARC researchers weren't Pollyannas. They also expressed misgivings about the world they foresaw. They worried that a ubiquitous computing system would be an ideal place for Big Brother to hide. "If the computational system is invisible as well as extensive," the lab's chief technologist, Mark Weiser, wrote in a 1999 article in *IBM*

Systems Journal, "it becomes hard to know what is controlling what, what is connected to what, where information is flowing, [and] how it is being used."[13] We'd have to place a whole lot of trust in the people and companies running the system.

The excitement about ubiquitous computing proved premature, as did the anxiety. The technology of the 1990s was not up to making the world machine-readable, and after the dot-com crash, investors were in no mood to bankroll the installation of expensive microchips and sensors everywhere. But much has changed in the succeeding fifteen years. The economic equations are different now. The price of computing gear has fallen sharply, as has the cost of high-speed data transmission. Companies like Amazon, Google, and Microsoft have turned data processing into a utility. They've built a cloud-computing grid that allows vast amounts of information to be collected and processed at efficient centralized plants and then fed into applications running on smartphones and tablets or into the control circuits of machines.[14] Manufacturers are spending billions of dollars to outfit factories with network-connected sensors, and technology giants like GE, IBM, and Cisco, hoping to spearhead the creation of an "internet of things," are rushing to develop standards for sharing the resulting data. Computers are pretty much omnipresent now, and even the faintest of the world's twitches and tremblings are being recorded as streams of binary digits. We may not be encalmed, but we are data saturated. The PARC researchers are starting to look like prophets.

There's a big difference between a set of tools and an infrastructure. The Industrial Revolution gained its full force only after its operational assumptions were built into expansive systems and networks. The construction of the railroads in the middle of the nineteenth century enlarged the markets companies could serve, providing the impetus for mechanized mass production and ever larger economies of scale. The creation of the electric grid a few decades later opened the way for factory assembly lines and, by making all sorts of electrical appliances feasible and affordable, spurred consumerism and pushed industrialization into the home. These new networks of transport and power, together with the telegraph, telephone, and broadcasting systems that arose alongside them, gave society a different character. They altered the way people thought about work, entertainment, travel, education, even the organization of communities and families. They transformed the pace and texture of life in ways that went well beyond what steam-powered factory machines had done.

Thomas Hughes, in reviewing the consequences of the arrival of the electric grid in his book *Networks of Power*, described how first the engineering culture, then the business culture, and finally the general culture shaped themselves to the new system. "Men and institutions developed characteristics that suited them to the characteristics of the technology," he wrote. "And the systematic interaction of men, ideas, and institutions, both technical and nontechnical, led to the development of a supersystem—a sociotechnical one—with mass movement and direction." It was at this point that technological momentum took hold, both for the power industry and for the modes of production and living it supported. "The universal system gathered a conservative momentum. Its growth generally was steady, and change became a diversification of function."[15] Progress had found its groove.

We've reached a similar juncture in the history of automation. Society is adapting to the universal computing infrastructure—more quickly than it adapted to the electric grid—and a new status quo is taking shape. The assumptions underlying industrial operations and commercial relations have already changed. "Business processes that once took place among human beings are now being executed electronically," explains W. Brian Arthur, an economist and technology theorist at the Santa Fe Institute. "They are taking place in an unseen domain that is strictly digital."[16] As an example, he points to the process of moving a shipment of freight through Europe. A few years ago, this would have required a legion of clipboard-wielding agents. They'd log arrivals and departures, check manifests, perform inspections, sign and stamp authorizations, fill out and file paperwork, and send letters or make phone calls to a variety of other functionaries involved in coordinating or regulating international freight. Changing the shipment's routing would have involved laborious communications among representatives of various concerned parties—shippers, receivers, carriers, government agencies—and more piles of paperwork. Now, pieces of cargo carry radio-frequency identification tags. When a shipment passes through a port or other way station, scanners read the tags and pass the information along to computers. The computers relay the information to other computers, which in concert perform the necessary checks, provide the required authorizations, revise schedules as needed, and make sure all parties have current data on the shipment's status. If a new routing is required, it's generated automatically and the tags and related data repositories are updated.

Such automated and far-flung exchanges of information have become routine throughout the economy. Commerce is increasingly managed through, as Arthur puts it, "a huge conversation conducted entirely among machines."[17] To be in business is to have networked computers capable of taking part in that conversation. "You know you have built an excellent digital nervous system," Bill Gates tells executives, "when information flows through your organization as quickly and naturally as thought in a human being."[18] Any sizable company, if it wants to remain viable, has little choice but to automate and then automate some more. It has to redesign its work flows and its products to allow for ever greater computer monitoring and control, and it has to restrict the involvement of people in its supply and production processes. People, after all, can't keep up with computer chatter; they just slow down the conversation.

The science-fiction writer Arthur C. Clarke once asked, "Can the synthesis of man and machine ever be stable, or will the purely organic component become such a hindrance that it has to be discarded?"[19] In the business world at least, no stability in the division of work between human and computer seems in the offing. The prevailing methods of computerized communication and coordination pretty much ensure that the role of people will go on shrinking. We've designed a system that discards us. If technological unemployment worsens in the years ahead, it will be more a result of our new, subterranean infrastructure of automation than of any particular installation of robots in factories or decision-support applications in offices. The robots and applications are the visible flora of automation's deep, extensive, and implacably invasive root system.

That root system is also feeding automation's spread into the broader culture. From the provision of government services to the tending of friendships and familial ties, society is reshaping itself to fit the contours of the new computing infrastructure. The infrastructure orchestrates the instantaneous data exchanges that make fleets of self-driving cars and armies of killer robots possible. It provides the raw material for the predictive algorithms that inform the decisions of individuals and groups. It underpins the automation of classrooms, libraries, hospitals, shops, churches, and homes—places traditionally associated with the human touch. It allows the NSA and other spy agencies, as well as crime syndicates and nosy corporations, to conduct surveillance and espionage on an unprecedented scale. It's what has shunted so much of our public discourse and private conversation onto tiny screens. And it's what gives our various computing devices the ability to guide us through the day, offering a steady stream of personalized alerts, instructions, and advice.

Once again, men and institutions are developing characteristics that suit them to the characteristics of the prevailing technology. Industrialization didn't turn us into machines, and automation isn't going to turn us into automatons. We're not that simple. But automation's spread is making our lives more programmatic. We have fewer opportunities to demonstrate our own resourcefulness and ingenuity, to display the self-reliance that was once considered the mainstay of character. Unless we start having second thoughts about where we're heading, that trend will only accelerate.

■   ■   ■   ■

IT WAS a curious speech. The event was the 2013 TED conference, held in late February at the Long Beach Performing Arts Center near Los Angeles. The scruffy guy on stage, fidgeting uncomfortably and talking in a halting voice, was Sergey Brin, reputedly the more outgoing of Google's two founders. He was there to deliver a marketing pitch for Glass, the company's "head-mounted computer." After airing a brief promotional video, he launched into a scornful critique of the smartphone, a device that Google, with its Android system, had helped push into the mainstream. Pulling his own phone from his pocket, Brin looked at it with disdain. Using a smartphone is "kind of emasculating," he said. "You know, you're standing around there, and you're just like rubbing this featureless piece of glass." In addition to being "socially isolating," staring down at a screen weakens a person's sensory engagement with the physical world, he suggested. "Is this what you were meant to do with your body?"[20]

Having dispatched the smartphone, Brin went on to extol the benefits of Glass. The new device would provide a far superior "form factor" for personal computing, he said. By freeing people's hands and allowing them to keep their head up and eyes forward, it would reconnect them with their surroundings. They'd rejoin the world. It had other advantages too. By putting a computer screen permanently within view, the high-tech eyeglasses would allow Google, through its Google Now service and other tracking and personalization routines, to deliver pertinent information to people whenever the device sensed they required advice or assistance. The company would fulfill the greatest of its ambitions: to automate the flow of information into the mind. Forget the autocomplete functions of Google Suggest. With Glass on your brow, Brin said, echoing his colleague Ray Kurzweil, you would no longer have to search the web at all. You wouldn't have to formulate queries or sort through results or follow trails of links. "You'd just have information come to you as you needed it."[21] To the computer's omnipresence would be added omniscience.

Brin's awkward presentation earned him the ridicule of technology bloggers. Still, he had a point. Smartphones enchant, but they also enervate. The human brain is incapable of concentrating on two things at once. Every glance or swipe at a touchscreen draws us away from our immediate surroundings. With a smartphone in hand, we become a little ghostly, wavering between worlds. People have always been distractible, of course. Minds wander. Attention drifts. But we've never carried on our person a tool that so insistently captivates our senses and divides our attention. By connecting us to a symbolic elsewhere, the smartphone, as Brin implied, exiles us from the here and now. We lose the power of presence.

Brin's assurance that Glass would solve the problem was less convincing. No doubt there are times when having your hands free while consulting a computer or using a camera would be an advantage. But peering into a screen that floats in front of you requires no less an investment of attention than glancing at one held in your lap. It may require more. Research on pilots and drivers who use head-up displays reveals that when people look at text or graphics projected as an overlay on the environment, they become susceptible to "attentional tunneling." Their focus narrows, their eyes fix on the display, and they become oblivious to everything else going on in their field of view.[22] In one experiment, performed in a flight simulator, pilots using a head-up display during a landing took longer to see a large plane obstructing the runway than did pilots who had to glance down to check their instrument readings. Two of the pilots using the head-up display never even saw the plane sitting directly in front of them.[23] "Perception requires both your eyes and your mind," psychology professors Daniel Simons and Christopher Chabris explained in a 2013 article on the dangers of Glass, "and if your mind is engaged, you can fail to see something that would otherwise be utterly obvious."[24]

Glass's display is also, by design, hard to escape. Hovering above your eye, it's always at the ready, requiring but a glance to call into view. At least a phone can be stuffed into a pocket or handbag, or slipped into a car's cup holder. The fact that you interact with Glass through spoken words, head movements, hand gestures, and finger taps further tightens its claim on the mind and senses. As for the audio signals that announce incoming alerts and messages—sent, as Brin boasted in his TED talk, "right through the bones in your cranium"—they hardly seem less intrusive than the beeps and buzzes of a phone. However emasculating a smartphone may be, metaphorically speaking, a computer attached to your forehead promises to be worse.

Wearable computers, whether sported on the head like Google's Glass and Facebook's Oculus Rift or on the wrist like the Pebble smartwatch, are new, and their appeal remains unproven. They'll have to overcome some big obstacles if they're to gain wide popularity. Their features are at this point sparse, they look dorky—London's *Guardian* newspaper refers to Glass as "those dreadful specs"[25]—and their tiny built-in cameras make a lot of people nervous. But, like other personal computers before them, they'll improve quickly, and they'll almost certainly morph into less obtrusive, more useful forms. The idea of wearing a computer may seem strange today, but in ten years it could be the norm. We may even find ourselves swallowing pill-sized nanocomputers to monitor our biochemistry and organ function.

Brin is mistaken, though, in suggesting that Glass and other such devices represent a break from computing's past. They give the established technological momentum even more force. As the smartphone and then the tablet made general-purpose, networked computers more portable and personable, they also made it possible for software companies to program many more aspects of our lives. Together with cheap, friendly apps, they allowed the cloud-computing infrastructure to be used to automate even the most mundane of chores. Computerized glasses and wristwatches further extend automation's reach. They make it easier to receive turn-by-turn directions when walking or riding a bike, for instance, or to get algorithmically generated advice on where to grab your next meal or what clothes to put on for a night out. They also serve as sensors for the body, allowing information about your location, thoughts, and health to be transmitted back to the cloud. That in turn provides software writers and entrepreneurs with yet more opportunities to automate the quotidian.

■   ■   ■   ■

WE'VE PUT into motion a cycle that, depending on your point of view, is either virtuous or vicious. As we grow more reliant on applications and algorithms, we become less capable of acting without their aid—we experience skill tunneling as well as attentional tunneling. That makes the software more indispensable still. Automation breeds automation. With everyone expecting to manage their lives through screens, society naturally adapts its routines and procedures to fit the routines and procedures of the computer. What can't be accomplished with software—what isn't amenable to computation and hence resists automation—begins to seem dispensable.

The PARC researchers argued, back in the early 1990s, that we'd know computing had achieved ubiquity when we were no longer aware of its presence. Computers would be so thoroughly enmeshed in our lives that they'd be invisible to us. We'd "use them unconsciously to accomplish everyday tasks."[26] That seemed a pipe dream in the days when bulky PCs drew attention to themselves by freezing, crashing, or otherwise misbehaving at inopportune moments. It doesn't seem like such a pipe dream anymore. Many computer companies and software houses now say they're working to make their products invisible. "I am super excited about technologies that disappear completely," declares Jack Dorsey, a prominent Silicon Valley entrepreneur. "We're doing this with Twitter, and we're doing this with [the online credit-card processor] Square."[27] When Mark Zuckerberg calls Facebook "a utility," as he frequently does, he's signaling that he wants the social network to merge into our lives the way the telephone system and electric grid did.[28] Apple has promoted the iPad as a device that "gets out of the way." Picking up on the theme, Google markets Glass as a means of "getting technology out of the way." In a 2013 speech, the company's then head of social networking, Vic Gundotra, even put a flower-power spin on the slogan: "Technology should get out of the way so you can live, learn, and love."[29]

The technologists may be guilty of bombast, but they're not guilty of cynicism. They're genuine in their belief that the more computerized our lives become, the happier we'll be. That, after all, has been their own experience. But their aspiration is self-serving nonetheless. For a popular technology to become invisible, it first has to become so essential to people's existence that they can no longer imagine being without it. It's only when a technology surrounds us that it disappears from view. Justin Rattner, Intel's chief technology officer, has said that he expects his company's products to become so much a part of people's "context" that Intel will be able to provide them with "pervasive assistance."[30] Instilling such dependency in customers would also, it seems safe to say, bring in a lot more money for Intel and other computer companies. For a business, there's nothing like turning a customer into a supplicant.

The prospect of having a complicated technology fade into the background, so it can be employed with little effort or thought, can be as appealing to those who use it as to those who sell it. "When technology gets out of the way, we are liberated from it," the *New York Times* columnist Nick Bilton has written.[31] But it's not that simple. You don't just flip a switch to make a technology invisible. It disappears only after a slow process of cultural and personal acclimation. As we habituate ourselves to it, the technology comes to exert more power over us, not less. We may be oblivious to the constraints it imposes on our lives, but the constraints remain. As the French sociologist Bruno Latour points out, the invisibility of a familiar technology is "a kind of optical illusion." It obscures the way we've refashioned ourselves to accommodate the technology. The tool that we originally used to fulfill some particular intention of our own begins to impose on us its intentions, or the intentions of its maker. "If we fail to recognize," Latour writes, "how much the use of a technique, however simple, has displaced, translated, modified, or inflected the initial intention, it is simply because we have *changed the end in changing the means*, and because, through a slipping of the will, we have begun to wish something quite else from what we at first desired."[32]

The difficult ethical questions raised by the prospect of programming robotic cars and soldiers—who controls the software? who chooses what's to be optimized? whose intentions and interests are reflected in the code?—are equally pertinent to the development of the applications used to automate our lives. As the programs gain more sway over us—shaping the way we work, the information we see, the routes we travel, our interactions with others—they become a form of remote control. Unlike robots or drones, we have the freedom to reject the software's instructions and suggestions. It's difficult, though, to escape their influence. When we launch an app, we ask to be guided—we place ourselves in the machine's care.

Look more closely at Google Maps. When you're traveling through a city and you consult the app, it gives you more than navigational tips; it gives you a way to think about cities. Embedded in the software is a philosophy of place, which reflects, among other things, Google's commercial interests, the backgrounds and biases of its programmers, and the strengths and limitations of software in representing space. In 2013, the company rolled out a new version of Google Maps. Instead of providing you with the same representation of a city that everyone else sees, it generates a map that's tailored to what Google perceives as your needs and desires, based on information the company has collected about you. The app will highlight nearby restaurants and other points of interest that friends in your social network have recommended. It will give you directions that reflect your past navigational choices. The views you see, the company says, are "unique to you, always adapting to the task you want to perform right this minute."[33]

That sounds appealing, but it's limiting. Google filters out serendipity in favor of insularity. It douses the infectious messiness of a city with an algorithmic antiseptic. What is arguably the most important way of looking at a city, as a public space shared not just with your pals but with an enormously varied group of strangers, gets lost. "Google's urbanism," comments the technology critic Evgeny Morozov, "is that of someone who is trying to get to a shopping mall in their self-driving car. It's profoundly utilitarian, even selfish in character, with little to no concern for how public space is experienced. In Google's world, public space is just something that stands between your house and the well-reviewed restaurant that you are dying to get to."[34] Expedience trumps all.

Social networks push us to present ourselves in ways that conform to the interests and prejudices of the companies that run them. Facebook, through its Timeline and other documentary features, encourages its members to think of their public image as indistinguishable from their identity. It wants to lock them into a single, uniform "self" that persists throughout their lives, unfolding in a coherent narrative beginning in childhood and ending, one presumes, with death. This fits with its founder's narrow conception of the self and its possibilities. "You have one identity," Mark Zuckerberg has said. "The days of you having a different image for your work friends or co-workers and for the other people you know are probably coming to an end pretty quickly." He even argues that "having two identities for yourself is an example of a lack of integrity."[35] That view, not surprisingly, dovetails with Facebook's desire to package its members as neat and coherent sets of data for advertisers. It has the added benefit, for the company, of making concerns about personal privacy seem less valid. If having more than one identity indicates a lack of integrity, then a yearning to keep certain thoughts or activities out of public view suggests a weakness of character. But the conception of selfhood that Facebook imposes through its software can be stifling. The self is rarely fixed. It has a protean quality. It emerges through personal exploration, and it shifts with circumstances. That's especially true in youth, when a person's self-conception is fluid, subject to testing, experimentation, and revision. To be locked into an identity, particularly early in one's life, may foreclose opportunities for personal growth and fulfillment.

Every piece of software contains such hidden assumptions. Search engines, in automating intellectual inquiry, give precedence to popularity and recency over diversity of opinion, rigor of argument, or quality of expression. Like all analytical programs, they have a bias toward criteria that lend themselves to statistical analysis, downplaying those that entail the exercise of taste or other subjective judgments. Automated essay-grading algorithms encourage in students a rote mastery of the mechanics of writing. The programs are deaf to tone, uninterested in knowledge's nuances, and actively resistant to creative expression. The deliberate breaking of a grammatical rule may delight a reader, but it's anathema to a computer. Recommendation engines, whether suggesting a movie or a potential love interest, cater to our established desires rather than challenging us with the new and unexpected. They assume we prefer custom to adventure, predictability to whimsy. The technologies of home automation, which allow things like lighting, heating, cooking, and entertainment to be meticulously programmed, impose a Taylorist mentality on domestic life. They subtly encourage people to adapt themselves to established routines and schedules, making homes more like workplaces.

The biases in software can distort societal decisions as well as personal ones. In promoting its self-driving cars, Google has suggested that the vehicles will dramatically reduce the number of crashes, if not eliminate them entirely. "Do you know that driving accidents are the number one cause of death for young people?" Sebastian Thrun said in a 2011 speech. "And do you realize that almost all of those are due to human error and not machine error, and can therefore be prevented by machines?"[36] Thrun's argument is compelling. In regulating hazardous activities like driving, society has long given safety a high priority, and everyone appreciates the role technological innovation can play in reducing the risk of mishaps and injuries. Even here, though, things aren't as black-and-white as Thrun implies. The ability of autonomous cars to prevent accidents and deaths remains theoretical at this point. As we've seen, the relationship between machinery and human error is complicated; it rarely plays out as expected. Society's goals, moreover, are never one-dimensional. Even the desire for safety requires interrogation. We've always recognized that laws and behavioral norms entail trade-offs between safety and liberty, between protecting ourselves and putting ourselves at risk. We allow and sometimes encourage people to engage in dangerous hobbies, sports, and other pursuits. A full life, we know, is not a perfectly insulated life. Even when it comes to setting speed limits on highways, we balance the goal of safety with other aims.

Difficult and often politically contentious, such trade-offs shape the kind of society we live in. The question is, do we want to cede the choices to software companies? When we look to automation as a panacea for human failings, we foreclose other options. A rush to embrace autonomous cars might do more than curtail personal freedom and responsibility; it might preclude us from exploring alternative ways to reduce the probability of traffic accidents, such as strengthening driver education or promoting mass transit.

It's worth noting that Silicon Valley's concern with highway safety, though no doubt sincere, has been selective. The distractions caused by cell phones and smartphones have in recent years become a major factor in car crashes. An analysis by the National Safety Council implicated phone use in one-fourth of all accidents on U.S. roads in 2012.[37] Yet Google and other top tech firms have made little or no effort to develop software to prevent people from calling, texting, or using apps while driving—surely a modest undertaking compared with building a car that can drive itself. Google has even sent its lobbyists into state capitals to block bills that would ban drivers from wearing Glass and other distracting eyewear. We should welcome the important contributions computer companies can make to society's well-being, but we shouldn't confuse those companies' interests with our own.

■   ■   ■   ■

IF WE don't understand the commercial, political, intellectual, and ethical motivations of the people writing our software, or the limitations inherent in automated data processing, we open ourselves to manipulation. We risk, as Latour suggests, replacing our own intentions with those of others, without even realizing that the swap has occurred. The more we habituate ourselves to the technology, the greater the risk grows.

It's one thing for indoor plumbing to become invisible, to fade from our view as we adapt ourselves, happily, to its presence. Even if we're incapable of fixing a leaky faucet or troubleshooting a balky toilet, we tend to have a pretty good sense of what the pipes in our homes do—and why. Most technologies that have become invisible to us through their ubiquity are like that. Their workings, and the assumptions and interests underlying their workings, are self-evident, or at least discernible. The technologies may have unintended effects— indoor plumbing changed the way people think about hygiene and privacy[38]—but they rarely have hidden agendas.

It's a very different thing for information technologies to become invisible. Even when we're conscious of their presence in our lives, computer systems are opaque to us. Software codes are hidden from our eyes, legally protected as trade secrets in many cases. Even if we could see them, few of us would be able to make sense of them. They're written in languages we don't understand. The data fed into algorithms is also concealed from us, often stored in distant, tightly guarded data centers. We have little knowledge of how the data is collected, what it's used for, or who has access to it. Now that software and data are stored in the cloud, rather than on personal hard drives, we can't even be sure when the workings of systems have changed. Revisions to popular programs are made all the time without our awareness. The application we used yesterday is probably not the application we use today.

The modern world has always been complicated. Fragmented into specialized domains of skill and knowledge, coiled with economic and other systems, it rebuffs any attempt to comprehend it in its entirety. But now, to a degree far beyond anything we've experienced before, the complexity itself is hidden from us. It's veiled behind the artfully contrived simplicity of the screen, the user-friendly, frictionless interface. We're surrounded by what the political scientist Langdon Winner has termed "concealed electronic complexity." The "relationships and connections" that were "once part of mundane experience," manifest in direct interactions among people and between people and things, have become "enshrouded in abstraction."[39] When an inscrutable technology becomes an invisible technology, we would be wise to be concerned. At that point, the technology's assumptions and intentions have infiltrated our own desires and actions. We no longer know whether the software is aiding us or controlling us. We're behind the wheel, but we can't be sure who's driving.

CHAPTER NINE

THE LOVE THAT LAYS THE SWALE IN ROWS

THERE'S A LINE OF VERSE I'M ALWAYS COMING BACK TO, and it's been on my mind even more than usual as I've worked my way through the manuscript of this book:

The fact is the sweetest dream that labor knows.

It's the second to last line of one of Robert Frost's earliest and best poems, a sonnet called "Mowing." He wrote it just after the turn of the twentieth century, when he was a young man, in his twenties, with a young family. He was working as a farmer, raising chickens and tending a few apple trees on a small plot of land his grandfather had bought for him in Derry, New Hampshire. It was a difficult time in his life. He had little money and few prospects. He had dropped out of two colleges, Dartmouth and Harvard, without earning a degree. He had been unsuccessful in a succession of petty jobs. He was sickly. He had nightmares. His firstborn child, a son, had died of cholera at the age of three. His marriage was troubled. "Life was peremptory," Frost would later recall, "and threw me into confusion."[1]

But it was during those lonely years in Derry that he came into his own as a writer and an artist. Something about farming—the long, repetitive days, the solitary work, the closeness to nature's beauty and carelessness—inspired him. The burden of labor eased the burden of life. "If I feel timeless and immortal it is from having lost track of time for five or six years there," he would write of his stay in Derry. "We gave up winding clocks. Our ideas got untimely from not taking newspapers for a long period. It couldn't have been more perfect if we had planned it or foreseen what we were getting into."[2] In the breaks between chores on the farm, Frost somehow managed to write most of the poems for his first book, *A Boy's Will*; about half the poems for his second book, *North of Boston*; and a good number of other poems that would find their way into subsequent volumes.

"Mowing," from *A Boy's Will*, was the greatest of his Derry lyrics. It was the poem in which he found his distinctive voice: plainspoken and conversational, but also sly and dissembling. (To really understand Frost—to really understand anything, including yourself—requires as much mistrust as trust.) As with many of his best works, "Mowing" has an enigmatic, almost hallucinatory quality that belies the simple and homely picture it paints—in this case of a man cutting a field of grass for hay. The more you read the poem, the deeper and stranger it becomes:

> There was never a sound beside the wood but one,
> And that was my long scythe whispering to the ground.
> What was it it whispered? I knew not well myself;
> Perhaps it was something about the heat of the sun,
> Something, perhaps, about the lack of sound—
> And that was why it whispered and did not speak.
> It was no dream of the gift of idle hours,

Or easy gold at the hand of fay or elf:
Anything more than the truth would have seemed too weak
To the earnest love that laid the swale in rows,
Not without feeble-pointed spikes of flowers
(Pale orchises), and scared a bright green snake.
The fact is the sweetest dream that labor knows.
My long scythe whispered and left the hay to make.[3]

We rarely look to poetry for instruction anymore, but here we see how a poet's scrutiny of the world can be more subtle and discerning than a scientist's. Frost understood the meaning of what we now call "flow" and the essence of what we now call "embodied cognition" long before psychologists and neurobiologists delivered the empirical evidence. His mower is not an airbrushed peasant, a romantic caricature. He's a farmer, a man doing a hard job on a still, hot summer day. He's not dreaming of "idle hours" or "easy gold." His mind is on his work—the bodily rhythm of the cutting, the weight of the tool in his hands, the stalks piling up around him. He's not seeking some greater truth beyond the work. The work is the truth.

The fact is the sweetest dream that labor knows.

There are mysteries in that line. Its power lies in its refusal to mean anything more or less than what it says. But it seems clear that what Frost is getting at, in the line and in the poem, is the centrality of action to both living and knowing. Only through work that brings us into the world do we approach a true understanding of existence, of "the fact." It's not an understanding that can be put into words. It can't be made explicit. It's nothing more than a whisper. To hear it, you need to get very near its source. Labor, whether of the body or the mind, is more than a way of getting things done. It's a form of contemplation, a way of seeing the world face-to-face rather than through a glass. Action un-mediates perception, gets us close to the thing itself. It binds us to the earth, Frost implies, as love binds us to one another. The antithesis of transcendence, work puts us in our place.

Frost is a poet of labor. He's always coming back to those revelatory moments when the active self blurs into the surrounding world—when, as he would write so memorably in another poem, "the work is play for mortal stakes."[4] The literary critic Richard Poirier, in his book *Robert Frost: The Work of Knowing*, described with great sensitivity the poet's view of the essence and essentialness of hard work: "Any intense labor enacted in his poetry, like mowing or apple-picking, can penetrate to the visions, dreams, myths that are at the heart of reality, constituting its articulate form for those who can read it with a requisite lack of certainty and an indifference to merely practical possessiveness." The knowledge gained through such efforts may be as shadowy and elusive as a dream—the very opposite of algorithmic or computational—but "in its mythic propensities, the knowledge is less ephemeral than are the apparently more practical results of labor, like food or money."[5] When we embark on a task, with our bodies or our minds, on our own or alongside others, we usually have a practical goal in sight. Our eyes are looking ahead to the product of our work—a store of hay for feeding livestock, perhaps. But it's through the work itself that we come to a deeper understanding of ourselves and our situation. The mowing, not the hay, is what matters most.

■   ■   ■   ■

NONE OF this should be taken as an attack on or a rejection of material progress. Frost is not romanticizing some distant, pre-technological past. Although he was dismayed by those who allowed themselves to become "bigoted in reliance / On the gospel of modern science,"[6] he felt a close kinship with scientists and inventors. As a poet, he shared with them a common spirit and a common pursuit. They were all explorers of the mysteries of earthly life, excavators of meaning from matter. They were all engaged in work that, as Poirier described it, "can extend the capability of human dreaming."[7] For Frost, the greatest value of "the fact"—whether apprehended in the world or expressed in a work of art or made manifest in a tool or other invention—lay in its ability to expand the scope of individual knowing and hence open new avenues of perception, action, and imagination. In the long poem "Kitty Hawk," written near the end of his life, he celebrated the Wright brothers' flight "Into the unknown, / Into the sublime." In making their own "pass / At the infinite," the brothers also made the experience of flight, and the sense of unboundedness it provides, possible for all of us. Theirs was a Promethean venture. In a sense, wrote Frost, the Wrights made the infinite "Rationally ours."[8]

Technology is as crucial to the work of knowing as it is to the work of production. The human body, in its native, unadorned state, is a feeble thing. It's constrained in its strength, its dexterity, its sensory range, its calculative prowess, its memory. It quickly reaches the limits of what it can do. But the body encompasses a mind that can imagine, desire, and plan for achievements the body alone can't fulfill. This tension between what the body can accomplish and what the mind can envision is what gave rise to and continues to propel and shape technology. It's the spur for humankind's extension of itself and elaboration of nature. Technology isn't what makes us "posthuman" or "transhuman," as some writers and scholars have recently suggested. It's what makes us human. Technology is in our nature. Through our tools we give our dreams form. We bring them into the world. The practicality of technology may distinguish it from art, but both spring from a similar, distinctly human yearning.

One of the many jobs the human body is unsuited to is cutting grass. (Try it if you don't believe me.) What allows the mower to do his work, what allows him to be a mower, is the tool he wields, his scythe. The mower is, and has to be, technologically enhanced. The tool makes the mower, and the mower's skill in using the tool remakes the world for him. The world becomes a place in which he can act as a mower, in which he can lay the swale in rows. This idea, which on the surface may sound trivial or even tautological, points to something elemental about life and the formation of the self.

"The body is our general means of having a world," wrote the French philosopher Maurice Merleau-Ponty in his 1945 masterwork *Phenomenology of Perception*.[9] Our physical makeup—the fact that we walk upright on two legs at a certain height, that we have a pair of hands with opposable thumbs, that we have eyes which see in a particular way, that we have a certain tolerance for heat and cold— determines our perception of the world in a way that precedes, and then molds, our conscious thoughts about the world. We see mountains as lofty not because mountains are lofty but because our perception of their form and height is shaped by our own stature. We see a stone as, among other things, a weapon because the particular construction of our hand and arm enables us to pick it up and throw it. Perception, like cognition, is embodied.

It follows that whenever we gain a new talent, we not only change our bodily capacities, we change the world. The ocean extends an invitation to the swimmer that it withholds from the person who has never learned to swim. With every skill we master, the world reshapes itself to reveal greater possibilities. It becomes more interesting, and being in it becomes more rewarding. This may be what Spinoza, the seventeenth-century Dutch philosopher who rebelled against Descartes' division of mind and body, was getting at when he wrote, "The human mind is capable of perceiving a great many things, and is the more capable, the more its body can be disposed in a great many ways."[10] John Edward Huth, a physics professor at Harvard, testifies to the regeneration that attends the mastery of a skill. A decade ago, inspired by Inuit hunters and other experts in natural wayfinding, he undertook "a self-imposed program to learn navigation through environmental clues." Through months of rigorous outdoor observation and practice, he taught himself how to read the nighttime and daytime skies, interpret the movements of clouds and waves, decipher the shadows cast by trees. "After a year of this endeavor," he recalls, "something dawned on me: the way I viewed the world had palpably changed. The sun looked different, as did the stars." Huth's enriched perception of the environment, gained through a kind of "primal empiricism," struck him as being "akin to what people describe as spiritual awakenings."[11]

Technology, by enabling us to act in ways that go beyond our bodily limits, also alters our perception of the world and what the world signifies to us. Technology's transformative power is most apparent in tools of discovery, from the microscope and the particle accelerator of the scientist to the canoe and the spaceship of the explorer, but the power is there in all tools, including the ones we use in our everyday lives. Whenever an instrument allows us to cultivate a new talent, the world becomes a different and more intriguing place, a setting of even greater opportunity. To the possibilities of nature are added the possibilities of culture. "Sometimes," wrote Merleau-Ponty, "the signification aimed at cannot be reached by the natural means of the body. We must, then, construct an instrument, and the body projects a cultural world around itself."[12] The value of a well-made and well-used tool lies not only in what it produces for us but what it produces in us. At its best, technology opens fresh ground. It gives us a world that is at once more understandable to our senses and better suited to our intentions—a world in which we're more at home. "My body is geared into the world when my perception provides me with the most varied and the most clearly articulated spectacle possible," explained Merleau-Ponty, "and when my motor intentions, as they unfold, receive the responses they anticipate from the world. This maximum of clarity in perception and action specifies a perceptual *ground*, a background for my life, a general milieu for the coexistence of my body and the world."[13] Used thoughtfully and with skill, technology becomes much more than a means of production or consumption. It becomes a means of experience. It gives us more ways to lead rich and engaged lives.

Look more closely at the scythe. It's a simple tool, but an ingenious one. Invented around 500 BC, by the Romans or the Gauls, it consists of a curved blade, forged of iron or steel, attached to the end of a long wooden pole, or snath. The snath typically has, about halfway down its length, a small wooden grip, or nib, that makes it possible to grasp and swing the implement with two hands. The scythe is a variation on the much-older sickle, a similar but short-handled cutting tool, invented in the Stone Age, that came to play an essential role in the early development of agriculture and, in turn, of civilization. What made the scythe a momentous innovation in its own right is that its long snath allowed a farmer or other laborer to cut grass at ground level while standing upright. Hay or grain could be harvested, or a pasture cleared, more quickly than before. Agriculture leaped forward.

The scythe enhanced the productivity of the worker in the field, but its benefit went beyond what could be measured in yield. The scythe was a congenial tool, far better suited to the bodily work of mowing than the sickle had been. Rather than stooping or squatting, the farmer could walk with a natural gait and use both his hands, as well as the full strength of his torso, in his job. The scythe served as both an aid and an invitation to the skilled work it enabled. We see in its form a model for technology on a human scale, for tools that extend the productive capabilities of society without circumscribing the individual's scope of action and perception. Indeed, as Frost makes clear in "Mowing," the scythe intensifies its user's involvement with and apprehension of the world. The mower swinging a scythe does more, but he also knows more. Despite outward appearances, the scythe is a tool of the mind as well as the body.

Not all tools are so congenial. Some deter us from skilled action. The digital technologies of automation, rather than inviting us into the world and encouraging us to develop new talents that enlarge our perceptions and expand our possibilities, often have the opposite effect. They're designed to be disinviting. They pull us away from the world. That's a consequence not only of the prevailing technology-centered design practices that place ease and efficiency above all other concerns. It also reflects the fact that, in our personal lives, the computer has become a media device, its software painstakingly programmed to grab and hold our attention. As most people know from experience, the computer screen is intensely compelling, not only for the conveniences it offers but also for the many diversions it provides.[14] There's always something going on, and we can join in at any moment with the slightest of effort. Yet the screen, for all its enticements and stimulations, is an environment of sparseness—fast-moving, efficient, clean, but revealing only a shadow of the world.

That's true even of the most meticulously crafted simulations of space that we find in virtual-reality applications such as games, CAD models, three-dimensional maps, and the tools used by surgeons and others to control robots. Artificial renderings of space may provide stimulation to our eyes and to a lesser degree our ears, but they tend to starve our other senses—touch, smell, taste—and greatly restrict the movements of our bodies. A study of rodents, published in *Science* in 2013, indicated that the brain's place cells are much less active when animals make their way through computer-generated landscapes than when they navigate the real world.[15] "Half of the neurons just shut up," reported one of the researchers, UCLA neurophysicist Mayank Mehta. He believes that the drop-off in mental activity likely stems from the lack of "proximal cues"—environmental smells, sounds, and textures that provide clues to location—in digital simulations of space.[16] "A map is not the territory it represents," the Polish philosopher Alfred Korzybski famously remarked,[17] and a virtual rendering is not the territory it represents either. When we enter the glass cage, we're required to shed much of our body. That doesn't free us; it emaciates us.

The world in turn is made less meaningful. As we adapt to our streamlined environment, we render ourselves incapable of perceiving what the world offers its most ardent inhabitants. Like the young, satellite-guided Inuit, we travel blindfolded. The result is existential impoverishment, as nature and culture withdraw their invitations to act and to perceive. The self can only thrive, can only grow, when it encounters and overcomes "resistance from surroundings," wrote John Dewey. "An environment that was always and everywhere congenial to the straightaway execution of our impulses would set a term to growth as sure as one always hostile would irritate and destroy. Impulsion forever boosted on its forward way would run its course thoughtless, and dead to emotion."[18]

Ours may be a time of material comfort and technological wonder, but it's also a time of aimlessness and gloom. During the first decade of this century, the number of Americans taking prescription drugs to treat depression or anxiety rose by nearly a quarter. One in five adults now regularly takes such medications.[19] The suicide rate among middle-aged Americans increased by nearly 30 percent over the same ten years, according to a report from the Centers for Disease Control and Prevention.[20] More than 10 percent of American schoolchildren, and nearly 20 percent of high-school-age boys, have been given a diagnosis of attention deficit hyperactivity disorder, and two-thirds of that group take drugs like Ritalin and Adderall to treat the condition.[21] The reasons for our discontent are many and far from understood. But one of them may be that through the pursuit of a frictionless existence, we've succeeded in turning what Merleau-Ponty termed the ground of our lives into a barren place. Drugs that numb the nervous system provide a way to rein in our vital, animal sensorium, to shrink our being to a size that better suits our constricted environs.

■   ■   ■   ■

FROST'S SONNET also contains, as one of its many whispers, a warning about technology's ethical hazards. There's a brutality to the mower's scythe. It indiscriminately cuts down flowers—those tender, pale orchises—along with the stalks of grass.* It frightens innocent animals, like the bright green snake. If technology embodies our dreams, it also embodies other, less benign qualities in our makeup, such as our will to power and the arrogance and insensitivity that accompany it. Frost returns to this theme a little later in *A Boy's Will*, in a second lyric about cutting hay, "The Tuft of Flowers." The poem's narrator comes upon a freshly mown field and, while following the flight of a passing butterfly with his eyes, discovers in the midst of the cut grass a small cluster of flowers, "a leaping tongue of bloom" that "the scythe had spared":

> The mower in the dew had loved them thus,
> By leaving them to flourish, not for us,
>
> Nor yet to draw one thought of us to him,
> But from sheer morning gladness to the brim.[22]

Working with a tool is never just a practical matter, Frost is telling us, with characteristic delicacy. It always entails moral choices and

---

* The destructive potential of the scythe gains even greater symbolic resonance when one remembers that the orchise, a tuberous plant, derives its name from the Greek word for testicle, *orkhis*. Frost was well versed in classical languages and literature. He would also have been familiar with the popular image of the Grim Reaper and his scythe.

has moral consequences. It's up to us, as users and makers of tools, to humanize technology, to aim its cold blade wisely. That requires vigilance and care.

The scythe is still employed in subsistence farming in many parts of the world. But it has no place on the modern farm, the development of which, like the development of the modern factory, office, and home, has required ever more complex and efficient equipment. The threshing machine was invented in the 1780s, the mechanical reaper appeared around 1835, the baler came a few years after that, and the combine harvester began to be produced commercially toward the end of the nineteenth century. The pace of technological advance has only accelerated in the decades since, and today the trend is reaching its logical conclusion with the computerization of agriculture. The working of the soil, which Thomas Jefferson saw as the most vigorous and virtuous of occupations, is being offloaded almost entirely to machines. Farmhands are being replaced by "drone tractors" and other robotic systems that, using sensors, satellite signals, and software, plant seeds, fertilize and weed fields, harvest and package crops, and milk cows and tend other livestock.[23] In development are robo-shepherds that guide flocks through pastures. Even if scythes still whispered in the fields of the industrial farm, no one would be around to hear them.

The congeniality of hand tools encourages us to take responsibility for their use. Because we sense the tools as extensions of our bodies, parts of ourselves, we have little choice but to be intimately involved in the ethical choices they present. The scythe doesn't choose to slash or spare the flowers; the mower does. As we become more expert in the use of a tool, our sense of responsibility for it naturally strengthens. To the novice mower, a scythe may feel like a foreign object in the hands; to the accomplished mower, hands and scythe become one thing. Talent tightens the bond between an instrument and its user. This feeling of physical and ethical entanglement doesn't have to go away as technologies become more complex. In reporting on his historic solo flight across the Atlantic in 1927, Charles Lindbergh spoke of his plane and himself as if they were a single being: "*We* have made this flight across the ocean, not *I* or *it*."[24] The airplane was a complicated system encompassing many components, but to a skilled pilot it still had the intimate quality of a hand tool. The love that lays the swale in rows is also the love that parts the clouds for the stick-and-rudder man.

Automation weakens the bond between tool and user not because computer-controlled systems are complex but because they ask so little of us. They hide their workings in secret code. They resist any involvement of the operator beyond the bare minimum. They discourage the development of skillfulness in their use. Automation ends up having an anesthetizing effect. We no longer feel our tools as parts of ourselves. In a seminal 1960 paper called "Man-Computer Symbiosis," the psychologist and engineer J. C. R. Licklider described the shift in our relation to technology well. "In the man-machine systems of the past," he wrote, "the human operator supplied the initiative, the direction, the integration, and the criterion. The mechanical parts of the systems were mere extensions, first of the human arm, then of the human eye." The introduction of the computer changed all that. "'Mechanical extension' has given way to replacement of men, to automation, and the men who remain are there more to help than to be helped."[25] The more automated everything gets, the easier it becomes to see technology as a kind of implacable, alien force that lies beyond our control and influence. Attempting to alter the path of its development seems futile. We press the on switch and follow the programmed routine.

To adopt such a submissive posture, however understandable it may be, is to shirk our responsibility for managing progress. A robotic harvesting machine may have no one in the driver's seat, but it is every bit as much a product of conscious human thought as a humble scythe is. We may not incorporate the machine into our brain maps, as we do the hand tool, but on an ethical level the machine still operates as an extension of our will. Its intentions are our intentions. If a robot scares a bright green snake (or worse), we're still to blame. We shirk a deeper responsibility as well: that of overseeing the conditions for the construction of the self. As computer systems and software applications come to play an ever larger role in shaping our lives and the world, we have an obligation to be more, not less, involved in decisions about their design and use—before technological momentum forecloses our options. We should be careful about what we make.

If that sounds naive or hopeless, it's because we have been misled by a metaphor. We've defined our relation with technology not as that of body and limb or even that of sibling and sibling but as that of master and slave. The idea goes way back. It took hold at the dawn of Western philosophical thought, emerging first, as Langdon Winner has described, with the ancient Athenians.[26] Aristotle, in discussing the operation of households at the beginning of his *Politics*, argued that slaves and tools are essentially equivalent, the former acting as "animate instruments" and the latter as "inanimate instruments" in the service of the master of the house. If tools could somehow become animate, Aristotle posited, they would be able to substitute directly for the labor of slaves. "There is only one condition on which we can imagine managers not needing subordinates, and masters not needing slaves," he mused, anticipating the arrival of computer automation and even machine learning. "This condition would be that each [inanimate] instrument could do its own work, at the word of command or by intelligent anticipation." It would be "as if a shuttle should weave itself, and a plectrum should do its own harp-playing."[27]

The conception of tools as slaves has colored our thinking ever since. It informs society's recurring dream of emancipation from toil, the one that was voiced by Marx and Wilde and Keynes and that continues to find expression in the works of technophiles and technophobes alike. "Wilde was right," Evgeny Morozov, the technology critic, wrote in his 2013 book *To Save Everything, Click Here*: "mechanical slavery is the enabler of human liberation."[28] We'll all soon have "personal workbots" at our "beck and call," Kevin Kelly, the technology enthusiast, proclaimed in a *Wired* essay that same year. "They will do jobs we have been doing, and do them much better than we can." More than that, they will free us to discover "new tasks that expand who we are. They will let us focus on becoming more human than we were."[29] *Mother Jones*'s Kevin Drum, also writing in 2013, declared that "a robotic paradise of leisure and contemplation eventually awaits us." By 2040, he predicted, our super-smart, super-reliable, super-compliant computer slaves—"they never get tired, they're never ill-tempered, they never make mistakes"—will have rescued us from labor and delivered us into an upgraded Eden. "Our days are spent however we please, perhaps in study, perhaps playing video games. It's up to us."[30]

With its roles reversed, the metaphor also informs society's nightmares about technology. As we become dependent on our technological slaves, the thinking goes, we turn into slaves ourselves. From the eighteenth century on, social critics have routinely portrayed factory machinery as forcing workers into bondage. "Masses of labourers," wrote Marx and Engels in their *Communist Manifesto*, "are daily and hourly enslaved by the machine."[31] Today, people complain all the time about feeling like slaves to their appliances and gadgets. "Smart devices are sometimes empowering," observed *The Economist* in "Slaves to the Smartphone," an article published in 2012. "But for most people the servant has become the master."[32] More dramatically still, the idea of a robot uprising, in which computers with artificial intelligence transform themselves from our slaves to our masters, has for a century been a central theme in dystopian fantasies about the future. The very word *robot*, coined by a science-fiction writer in 1920, comes from *robota*, a Czech term for servitude.

The master-slave metaphor, in addition to being morally fraught, distorts the way we look at technology. It reinforces the sense that our tools are separate from ourselves, that our instruments have an agency independent of our own. We start to judge our technologies not on what they enable us to do but rather on their intrinsic qualities as products—their cleverness, their efficiency, their novelty, their style. We choose a tool because it's new or it's cool or it's fast, not because it brings us more fully into the world and expands the ground of our experiences and perceptions. We become mere consumers of technology.

More broadly, the metaphor encourages society to take a simplistic and fatalistic view of technology and progress. If we assume that our tools act as slaves on our behalf, always working in our best interest, then any attempt to place limits on technology becomes hard to defend. Each advance grants us greater freedom and takes us a stride closer to, if not utopia, then at least the best of all possible worlds. Any misstep, we tell ourselves, will be quickly corrected by subsequent innovations. If we just let progress do its thing, it will find remedies for the problems it creates. "Technology is not neutral but serves as an overwhelming positive force in human culture," writes Kelly, expressing the self-serving Silicon Valley ideology that in recent years has gained wide currency. "We have a moral obligation to increase technology because it increases opportunities."[33] The sense of moral obligation strengthens with the advance of automation, which, after all, provides us with the most animate of instruments, the slaves that, as Aristotle anticipated, are most capable of releasing us from our labors.

The belief in technology as a benevolent, self-healing, autonomous force is seductive. It allows us to feel optimistic about the future while relieving us of responsibility for that future. It particularly suits the interests of those who have become extraordinarily wealthy through the labor-saving, profit-concentrating effects of automated systems and the computers that control them. It provides our new plutocrats with a heroic narrative in which they play starring roles: recent job losses may be unfortunate, but they're a necessary evil on the path to the human race's eventual emancipation by the computerized slaves that our benevolent enterprises are creating. Peter Thiel, a successful entrepreneur and investor who has become one of Silicon Valley's most prominent thinkers, grants that "a robotics revolution would basically have the effect of people losing their jobs." But, he hastens to add, "it would have the benefit of freeing people up to do many other things."[34] Being freed up sounds a lot more pleasant than being fired.

There's a callousness to such grandiose futurism. As history reminds us, high-flown rhetoric about using technology to liberate workers often masks a contempt for labor. It strains credulity to imagine today's technology moguls, with their libertarian leanings and impatience with government, agreeing to the kind of vast wealth-redistribution scheme that would be necessary to fund the self-actualizing leisure-time pursuits of the jobless multitudes. Even if society were to come up with some magic spell, or magic algorithm, for equitably parceling out the spoils of automation, there's good reason to doubt whether anything resembling the "economic bliss" imagined by Keynes would ensue. In a prescient passage in *The Human Condition*, Hannah Arendt observed that if automation's utopian promise were actually to pan out, the result would probably feel less like paradise than like a cruel practical joke. The whole of modern society, she wrote, has been organized as "a laboring society," where working for pay, and then spending that pay, is the way people define themselves and measure their worth. Most of the "higher and more meaningful activities" revered in the distant past have been pushed to the margin or forgotten, and "only solitary individuals are left who consider what they are doing in terms of work and not in terms of making a living." For technology to fulfill humankind's abiding "wish to be liberated from labor's 'toil and trouble'" at this point would be perverse. It would cast us deeper into a purgatory of malaise. What automation confronts us with, Arendt concluded, "is the prospect of a society of laborers without labor, that is, without the only activity left to them. Surely, nothing could be worse."[35] Utopianism, she understood, is a form of miswanting.

The social and economic problems caused or exacerbated by automation aren't going to be solved by throwing more software at them. Our inanimate slaves aren't going to chauffeur us to a utopia of comfort and harmony. If the problems are to be solved, or at least attenuated, the public will need to grapple with them in their full complexity. To ensure society's well-being in the future, we may need to place limits on automation. We may have to shift our view of progress, putting the emphasis on social and personal flourishing rather than technological advancement. We may even have to entertain an idea that's come to be considered unthinkable, at least in business circles: giving people precedence over machines.

■　　■　　■　　■

In 1986, a Canadian ethnographer named Richard Kool wrote Mihaly Csikszentmihalyi a letter. Kool had read some of the professor's early work about flow, and he had been reminded of his own research into the Shushwap tribe, an aboriginal people who lived in the Thompson River Valley in what is now British Columbia. The Shushwap territory was "a plentiful land," Kool noted. It was blessed with an abundance of fish and game and edible roots and berries. The Shushwaps did not have to wander to survive. They built villages and developed "elaborate technologies for very effectively using the resources in the environment." They viewed their lives as good and rich. But the tribe's elders saw that in such comfortable circumstances lay danger. "The world became too predictable and the challenge began to go out of life. Without challenge, life had no meaning." And so, every thirty years or so, the Shushwaps, led by their elders, would uproot themselves. They'd leave their homes, abandon their villages, and head off into the wilds. "The entire population," Kool reported, "would move to a different part of the Shushwap land." And there they would discover a fresh set of challenges. "There were new streams to figure out, new game trails to learn, new areas where the balsam root would be plentiful. Now life would regain its meaning and be worth living. Everyone would feel rejuvenated and happy."[36]

■   ■   ■   ■

E. J. MEADE, the Colorado architect, said something revealing when I talked to him about his firm's adoption of computer-aided design systems. The hard part wasn't learning how to use the software. That was pretty easy. What was tough was learning how *not* to use it. The speed, ease, and sheer novelty of CAD made it enticing. The first instinct of the firm's designers was to plop themselves down at their computers at the start of a project. But when they took a hard look at their work, they realized that the software was a hindrance to creativity. It was closing off aesthetic and functional possibilities even as it was quickening the pace of production. As Meade and his colleagues thought more critically about the effects of automation, they began to resist the technology's temptations. They found themselves "bringing the computer in later and later" in the course of a project. For the early, formative stages of the work, they returned to their sketchbooks and sheets of tracing paper, their models of cardboard and foam core. "At the back end, it's brilliant," Meade said in summing up what he's learned about CAD. "The convenience factor is great." But the computer's "expedience" can be perilous. For the unwary and the uncritical, it can overwhelm other, more important considerations. "You have to dig deep into the tool to avoid being manipulated by it."

A year or so before I talked to Meade—just as I was beginning the research for this book—I had a chance meeting on a college campus with a freelance photographer who was working on an assignment for the school. He was standing idly under a tree, waiting for some uncooperative clouds to get out of the way of the sun. I noticed he had a large-format film camera set up on a bulky tripod—it was hard to miss, as it looked almost absurdly old-fashioned—and I asked him why he was still using film. He told me that he had eagerly embraced digital photography a few years earlier. He had replaced his film cameras and his darkroom with digital cameras and a computer running the latest image-processing software. But after a few months, he switched back. It wasn't that he was dissatisfied with the operation of the equipment or the resolution or accuracy of the images. It was that the way he went about his work had changed, and not for the better.

The constraints inherent in taking and developing pictures on film—the expense, the toil, the uncertainty—had encouraged him to work slowly when he was on a shoot, with deliberation, thoughtfulness, and a deep, physical sense of presence. Before he took a picture, he would compose the shot meticulously in his mind, attending to the scene's light, color, framing, and form. He would wait patiently for the right moment to release the shutter. With a digital camera, he could work faster. He could take a slew of images, one after the other, and then use his computer to sort through them and crop and tweak the most promising ones. The act of composition took place after a photo was taken. The change felt intoxicating at first. But he found himself disappointed with the results. The images left him cold. Film, he realized, imposed a discipline of perception, of seeing, which led to richer, more artful, more moving photographs. Film demanded more of him. And so he went back to the older technology.

Neither the architect nor the photographer was the least bit antagonistic toward computers. Neither was motivated by abstract concèrns about a loss of agency or autonomy. Neither was a crusader. Both just wanted the best tool for the job—the tool that would encourage and enable them to do their finest, most fulfilling work. What they came to realize was that the newest, most automated, most expedient tool is not always the best choice. Although I'm sure they would bristle at being likened to the Luddites, their decision to forgo the latest technology, at least in some stages of their work, was an act of rebellion resembling that of the old English machine-breakers, if without the fury and the violence. Like the Luddites, they understood that decisions about technology are also decisions about ways of working and ways of living—and they took control of those decisions rather than ceding them to others or giving way to the momentum of progress. They stepped back and thought critically about technology.

As a society, we've become suspicious of such acts. Out of ignorance or laziness or timidity, we've turned the Luddites into caricatures, emblems of backwardness. We assume that anyone who rejects a new tool in favor of an older one is guilty of nostalgia, of making choices sentimentally rather than rationally. But the real sentimental fallacy is the assumption that the new thing is always better suited to our purposes and intentions than the old thing. That's the view of a child, naive and pliable. What makes one tool superior to another has nothing to do with how new it is. What matters is how it enlarges us or diminishes us, how it shapes our experience of nature and culture and one another. To cede choices about the texture of our daily lives to a grand abstraction called progress is folly.

Technology has always challenged people to think about what's important in their lives, to ask themselves, as I suggested at the outset of this book, what *human being* means. Automation, as it extends its reach into the most intimate spheres of our existence, raises the stakes. We can allow ourselves to be carried along by the technological current, wherever it may be taking us, or we can push against it.

To resist invention is not to reject invention. It's to humble invention, to bring progress down to earth. "Resistance is futile," goes the glib *Star Trek* cliché beloved by techies. But that's the opposite of the truth. Resistance is never futile. If the source of our vitality is, as Emerson taught us, "the active soul,"[37] then our highest obligation is to resist any force, whether institutional or commercial or technological, that would enfeeble or enervate the soul.

One of the most remarkable things about us is also one of the easiest to overlook: each time we collide with the real, we deepen our understanding of the world and become more fully a part of it. While we're wrestling with a challenge, we may be motivated by an anticipation of the ends of our labor, but, as Frost saw, it's the work—the means—that makes us who we are. Automation severs ends from means. It makes getting what we want easier, but it distances us from the work of knowing. As we transform ourselves into creatures of the screen, we face the same existential question that the Shushwap confronted: Does our essence still lie in what we know, or are we now content to be defined by what we want?

That sounds very serious. But the aim is joy. The active soul is a light soul. By reclaiming our tools as parts of ourselves, as instruments of experience rather than just means of production, we can enjoy the freedom that congenial technology provides when it opens the world more fully to us. It's the freedom I imagine Lawrence Sperry and Emil Cachin must have felt on that bright spring day in Paris a hundred years ago when they climbed out onto the wings of their gyroscope-balanced Curtiss C-2 biplane and, filled with terror and delight, passed over the reviewing stands and saw below them the faces of the crowd turned skyward in awe.

NOTES

Introduction: ALERT FOR OPERATORS

1. Federal Aviation Administration, SAFO 13002, January 4, 2013, faa
   .gov/other_visit/aviation_industry/airline_operators/airline_safety/safo/
   all_safos/media/2013/SAFO13002.pdf.

Chapter One: PASSENGERS

1. Sebastian Thrun, "What We're Driving At," *Google Official Blog*, October 9, 2010, googleblog.blogspot.com/2010/10/what-were-driving-at
   .html. See also Tom Vanderbilt, "Let the Robot Drive: The Autonomous
   Car of the Future Is Here," *Wired*, February 2012.
2. Daniel DeBolt, "Google's Self-Driving Car in Five-Car Crash," *Moun-
   tain View Voice*, August 8, 2011.
3. Richard Waters and Henry Foy, "Tesla Moves Ahead of Google in
   Race to Build Self-Driving Cars," *Financial Times*, September 17, 2013,
   ft.com/intl/cms/s/0/70d26288-1faf-11e3-8861-00144feab7de.html.
4. Frank Levy and Richard J. Murnane, *The New Division of Labor: How
   Computers Are Creating the Next Job Market* (Princeton: Princeton University Press, 2004), 20.
5. Tom A. Schweizer et al., "Brain Activity during Driving with Distrac-
   tion: An Immersive fMRI Study," *Frontiers in Human Neuroscience*,

February 28, 2013, frontiersin.org/Human_Neuroscience/10.3389/fnhum.2013.00053/full.

6. N. Katherine Hayles, *How We Think: Digital Media and Contemporary Technogenesis* (Chicago: University of Chicago Press, 2012), 2.

7. Mihaly Csikszentmihalyi and Judith LeFevre, "Optimal Experience in Work and Leisure," *Journal of Personality and Social Psychology* 56, no. 5 (1989): 815–822.

8. Daniel T. Gilbert and Timothy D. Wilson, "Miswanting: Some Problems in the Forecasting of Future Affective States," in Joseph P. Forgas, ed., *Feeling and Thinking: The Role of Affect in Social Cognition* (Cambridge, U.K.: Cambridge University Press, 2000), 179.

9. Csikszentmihalyi and LeFevre, "Optimal Experience in Work and Leisure."

10. Quoted in John Geirland, "Go with the Flow," *Wired*, September 1996.

11. Mihaly Csikszentmihalyi, *Flow: The Psychology of Optimal Experience* (New York: Harper, 1991), 157–162.

Chapter Two: THE ROBOT AT THE GATE

1. R. H. Macmillan, *Automation: Friend or Foe?* (Cambridge, U.K.: Cambridge University Press, 1956), 1.

2. Ibid., 91.

3. Ibid., 1–6. The emphasis is Macmillan's.

4. Ibid., 92.

5. George B. Dyson, *Darwin among the Machines: The Evolution of Global Intelligence* (Reading, Mass.: Addison-Wesley, 1997), x.

6. Bertrand Russell, "Machines and the Emotions," in *Sceptical Essays* (London: Routledge, 2004), 64.

7. Adam Smith, *The Wealth of Nations* (New York: Modern Library, 2000), 7–10.

8. Ibid., 408.

9. Malcolm I. Thomis, *The Luddites: Machine-Breaking in Regency England* (Newton Abbot, U.K.: David & Charles, 1970), 50. See also E. J. Hobsbawm, "The Machine Breakers," *Past and Present* 1, no. 1 (1952): 57–70.

10. Karl Marx, *Capital: A Critique of Political Economy*, vol. 1 (Chicago: Charles H. Kerr, 1912), 461–462.

11. Karl Marx, "Speech at the Anniversary of the People's Paper," April 14, 1856, marxists.org/archive/marx/works/1856/04/14.htm.

12. Nick Dyer-Witheford, *Cyber-Marx: Cycles and Circuits of Struggle in High Technology Capitalism* (Champaign, Ill.: University of Illinois Press, 1999), 40.

13. Marx, "Speech at the Anniversary of the People's Paper."

14. Quoted in Dyer-Witheford, *Cyber-Marx*, 41. In a famous passage in *The German Ideology*, published in 1846, Marx foresaw a day when he would be free "to do one thing today and another tomorrow, to hunt in the morning, fish in the afternoon, rear cattle in the evening, criticise after dinner, just as I have a mind, without ever becoming hunter, fisherman, shepherd or critic." Miswanting has rarely sounded so rhapsodic.

15. E. Levasseur, "The Concentration of Industry, and Machinery in the United States," *Publications of the American Academy of Political and Social Science*, no. 193 (1897): 178–197.

16. Oscar Wilde, "The Soul of Man under Socialism," in *The Collected Works of Oscar Wilde* (Ware, U.K.: Wordsworth Editions, 2007), 1051.

17. Quoted in Amy Sue Bix, *Inventing Ourselves out of Jobs? America's Debate over Technological Unemployment, 1929–1981* (Baltimore: Johns Hopkins University Press, 2000), 117–118.

18. Ibid., 50.

19. Ibid., 55.

20. John Maynard Keynes, "Economic Possibilities for Our Grandchildren," in *Essays in Persuasion* (New York: W. W. Norton, 1963), 358–373.

21. John F. Kennedy, "Remarks at the Wheeling Stadium," in *John F. Kennedy: Containing the Public Messages, Speeches, and Statements of the President* (Washington, D.C.: U.S. Government Printing Office, 1962), 721.

22. Stanley Aronowitz and William DiFazio, *The Jobless Future: Sci-Tech and the Dogma of Work* (Minneapolis: University of Minnesota Press, 1994), 14. The emphasis is Aronowitz and DiFazio's.

23. Jeremy Rifkin, *The End of Work: The Decline of the Global Labor Force and the Dawn of the Post-Market Era* (New York: Putnam, 1995), xv–xviii.

24. Erik Brynjolfsson and Andrew McAfee, *Race against the Machine: How the Digital Revolution Is Accelerating Innovation, Driving Productivity, and Irreversibly Transforming Employment and the Economy* (Lexington, Mass.: Digital Frontier Press, 2011). Brynjolfsson and McAfee extended

their argument in *The Second Machine Age: Work, Progress, and Prosperity in a Time of Brilliant Technologies* (New York: W. W. Norton, 2014).

25. "March of the Machines," *60 Minutes*, CBS, January 13, 2013, cbsnews.com/8301-18560_162-57563618/are-robots-hurting-job-growth/.

26. Bernard Condon and Paul Wiseman, "Recession, Tech Kill Middle-Class Jobs," AP, January 23, 2013, bigstory.ap.org/article/ap-impact-recession-tech-kill-middle-class-jobs.

27. Paul Wiseman and Bernard Condon, "Will Smart Machines Create a World without Work?," AP, January 25, 2013, bigstory.ap.org/article/will-smart-machines-create-world-without-work.

28. Michael Spence, "Technology and the Unemployment Challenge," *Project Syndicate*, January 15, 2013, project-syndicate.org/commentary/global-supply-chains-on-the-move-by-michael-spence.

29. See Timothy Aeppel, "Man vs. Machine, a Jobless Recovery," *Wall Street Journal*, January 17, 2012.

30. Quoted in Thomas B. Edsall, "The Hollowing Out," *Campaign Stops* (blog), *New York Times*, July 8, 2012, campaignstops.blogs.nytimes.com/2012/07/08/the-future-of-joblessness/.

31. See Lawrence V. Kenton, ed., *Manufacturing Output, Productivity and Employment Implications* (New York: Nova Science, 2005); and Judith Banister and George Cook, "China's Employment and Compensation Costs in Manufacturing through 2008," *Monthly Labor Review*, March 2011.

32. Tyler Cowen, "What Export-Oriented America Means," *American Interest*, May/June 2012.

33. Robert Skidelsky, "The Rise of the Robots," *Project Syndicate*, February 19, 2013, project-syndicate.org/commentary/the-future-of-work-in-a-world-of-automation-by-robert-skidelsky.

34. Ibid.

35. Chrystia Freeland, "China, Technology and the U.S. Middle Class," *Financial Times*, February 15, 2013.

36. Paul Krugman, "Is Growth Over?," *The Conscience of a Liberal* (blog), *New York Times*, December 26, 2012, krugman.blogs.nytimes.com/2012/12/26/is-growth-over/.

37. James R. Bright, *Automation and Management* (Cambridge, Mass.: Harvard University, 1958), 4–5.

38. Ibid., 5.

39. Ibid., 4, 6. The emphasis is Bright's. Bright's definition of automation echoes Sigfried Giedion's earlier definition of mechanization: "Mechanization is an agent—like water, fire, light. It is blind and without direction of its own. Like the powers of nature, mechanization depends on man's capacity to make use of it and to protect himself from its inherent perils. Because mechanization sprang entirely from the mind of man, it is the more dangerous to him." Giedion, *Mechanization Takes Command* (New York: Oxford University Press, 1948), 714.

40. David A. Mindell, *Between Human and Machine: Feedback, Control, and Computing before Cybernetics* (Baltimore: Johns Hopkins University Press, 2002), 247.

41. Stuart Bennett, *A History of Control Engineering, 1800–1930* (London: Peter Peregrinus, 1979), 99–100.

42. Norbert Wiener, *The Human Use of Human Beings: Cybernetics and Society* (New York: Da Capo, 1954), 153.

43. Eric W. Leaver and J. J. Brown, "Machines without Men," *Fortune*, November 1946. See also David F. Noble, *Forces of Production: A Social History of Industrial Automation* (New York: Alfred A. Knopf, 1984), 67–71.

44. Noble, *Forces of Production*, 234.

45. Ibid., 21–40.

46. Wiener, *Human Use of Human Beings*, 148–162.

47. Quoted in Flo Conway and Jim Siegelman, *Dark Hero of the Information Age: In Search of Norbert Wiener, the Father of Cybernetics* (New York: Basic Books, 2005), 251.

48. Marc Andreessen, "Why Software Is Eating the World," *Wall Street Journal*, August 20, 2011.

Chapter Three: ON AUTOPILOT

1. The account of the Continental Connection crash draws primarily from the National Transportation Safety Board's Accident Report AAR-10/01: *Loss of Control on Approach, Colgan Air, Inc., Operating as Continental Connection Flight 3407, Bombardier DHC 8-400, N200WQ, Clarence, New York, February 12, 2009* (Washington, D.C.: NTSB, 2010), www.ntsb.gov/doclib/reports/2010/aar1001.pdf. See also Matthew L. Wald,

"Pilots Chatted in Moments before Buffalo Crash," *New York Times*, May 12, 2009.

2. Associated Press, "Inquiry in New York Air Crash Points to Crew Error," *Los Angeles Times*, May 13, 2009.

3. The account of the Air France crash draws primarily from BEA, *Final Report: On the Accident on 1st June 2009 to the Airbus A330-203, Registered F-GZCP, Operated by Air France, Flight AF447, Rio de Janeiro to Paris* (official English translation), July 27, 2012, www.bea.aero /docspa/2009/f-cp090601.en/pdf/f-cp090601.en.pdf. See also Jeff Wise, "What Really Happened Aboard Air France 447," *Popular Mechanics*, December 6, 2011, www.popularmechanics.com/technology/aviation /crashes/what-really-happened-aboard-air-france-447-6611877.

4. BEA, *Final Report*, 199.

5. William Scheck, "Lawrence Sperry: Genius on Autopilot," *Aviation History*, November 2004; Dave Higdon, "Used Correctly, Autopilots Offer Second-Pilot Safety Benefits," *Avionics News*, May 2010; and Anonymous, "George the Autopilot," *Historic Wings*, August 30, 2012, fly.his toricwings.com/2012/08/george-the-autopilot/.

6. "Now—The Automatic Pilot," *Popular Science Monthly*, February 1930.

7. "Post's Automatic Pilot," *New York Times*, July 24, 1933.

8. James M. Gillespie, "We Flew the Atlantic 'No Hands,'" *Popular Science*, December 1947.

9. Anonymous, "Automatic Control," *Flight*, October 9, 1947.

10. For a thorough account of NASA's work, see Lane E. Wallace, *Airborne Trailblazer: Two Decades with NASA Langley's 737 Flying Laboratory* (Washington, D.C.: NASA History Office, 1994).

11. William Langewiesche, *Fly by Wire: The Geese, the Glide, the "Miracle" on the Hudson* (New York: Farrar, Straus & Giroux, 2009), 103.

12. Antoine de Saint-Exupéry, *Wind, Sand and Stars* (New York: Reynal & Hitchcock, 1939), 20.

13. Don Harris, *Human Performance on the Flight Deck* (Surrey, U.K.: Ashgate, 2011), 221.

14. "How Does Automation Affect Airline Safety?," Flight Safety Foundation, July 3, 2012, flightsafety.org/node/4249.

15. Hemant Bhana, "Trust but Verify," *AeroSafety World*, June 2010.

16. Quoted in Nick A. Komons, *Bonfires to Beacons: Federal Civil Aviation Policy under the Air Commerce Act 1926–1938* (Washington, D.C.: U.S. Department of Transportation, 1978), 24.

17. Scott Mayerowitz and Joshua Freed, "Air Travel Safer than Ever with Death Rate at Record Low," Denverpost.com, January 1 , 2012, denver post.com/nationworld/ci_19653967. Deaths from terrorist acts are not included in the figures.

18. Interview of Raja Parasuraman by author, December 18, 2011.

19. Jan Noyes, "Automation and Decision Making," in Malcolm James Cook et al., eds., *Decision Making in Complex Environments* (Aldershot, U.K.: Ashgate, 2007), 73.

20. Earl L. Wiener, *Human Factors of Advanced Technology ("Glass Cockpit") Transport Aircraft* (Moffett Field, Calif.: NASA Ames Research Center, June 1989).

21. See, for example, Earl L. Wiener and Renwick E. Curry, "Flight-Deck Automation: Promises and Problems," NASA Ames Research Center, June 1980; Earl L. Wiener, "Beyond the Sterile Cockpit," *Human Factors* 27, no. 1 (1985): 75–90; Donald Eldredge et al., *A Review and Discussion of Flight Management System Incidents Reported to the Aviation Safety Reporting System* (Washington, D.C.: Federal Aviation Administration, February 1992); and Matt Ebbatson, "Practice Makes Imperfect: Common Factors in Recent Manual Approach Incidents," *Human Factors and Aerospace Safety* 6, no. 3 (2006): 275–278.

22. Andy Pasztor, "Pilot Reliance on Automation Erodes Skills," *Wall Street Journal*, November 5, 2010.

23. *Operational Use of Flight Path Management Systems: Final Report of the Performance-Based Operations Aviation Rulemaking Committee/Commercial Aviation Safety Team Flight Deck Automation Working Group* (Washington, D.C.: Federal Aviation Administration, September 5, 2013), www.faa.gov/about/office_org/headquarters_offices/avs/offices/afs /afs400/parc/parc_reco/media/2013/130908_PARC_FltDAWG _Final_Report_Recommendations.pdf.

24. Matthew Ebbatson, "The Loss of Manual Flying Skills in Pilots of Highly Automated Airliners" (PhD thesis, Cranfield University School of Engineering, 2009). See also M. Ebbatson et al., "The Relationship between Manual Handling Performance and Recent Flying Experience in Air Transport Pilots," *Ergonomics* 53, no. 2 (2010): 268–277.

25. Quoted in David A. Mindell, *Between Human and Machine: Feedback, Control, and Computing before Cybernetics* (Baltimore: Johns Hopkins University Press, 2002), 77.

26. S. Bennett, *A History of Control Engineering, 1800–1930* (Stevenage, U.K.: Peter Peregrinus, 1979), 141.

27. Tom Wolfe, *The Right Stuff* (New York: Picador, 1979), 152–154.

28. Ebbatson, "Loss of Manual Flying Skills."

29. European Aviation Safety Agency, "Response Charts for 'EASA Cockpit Automation Survey,'" August 3, 2012, easa.europa.eu/safety-and-research/docs/EASA%20Cockpit%20Automation%20Survey%20 2012%20-%20Results.pdf.

30. Joan Lowy, "Automation in the Air Dulls Pilot Skill," *Seattle Times*, August 30, 2011.

31. For a good review of changes in the size of flight crews, see Delmar M. Fadden et al., "First Hand: Evolution of the 2-Person Crew Jet Transport Flight Deck," *IEEE Global History Network*, August 25, 2008, ieeeghn .org/wiki/index.php/First-Hand:Evolution_of_the_2-Person_Crew_ Jet_Transport_Flight_Deck.

32. Quoted in Philip E. Ross, "When Will We Have Unmanned Commercial Airliners?," *IEEE Spectrum*, December 2011.

33. Scott McCartney, "Pilot Pay: Want to Know How Much Your Captain Earns?," *The Middle Seat Terminal* (blog), *Wall Street Journal*, June 16, 2009, blogs.wsj.com/middleseat/2009/06/16/pilot-pay-want-to-know-how-much-your-captain-earns/.

34. Dawn Duggan, "The 8 Most Overpaid & Underpaid Jobs," Salary.com, undated, salary.com/the%2D8%2Dmost%2Doverpaid%2Dunderpaid% 2Djobs/slide/9/.

35. David A. Mindell, *Digital Apollo: Human and Machine in Spaceflight* (Cambridge, Mass.: MIT Press, 2011), 20.

36. Wilbur Wright, letter, May 13, 1900, in Richard Rhodes, ed., *Visions of Technology: A Century of Vital Debate about Machines, Systems, and the Human World* (New York: Touchstone, 1999), 33.

37. Mindell, *Digital Apollo*, 20.

38. Quoted in ibid., 21.

39. Wilbur Wright, "Some Aeronautical Experiments," speech before the Western Society of Engineers, September 18, 1901, www.wright-house .com/wright-brothers/Aeronautical.html.

40. Mindell, *Digital Apollo*, 21.

41. J. O. Roberts, "The Case against Automation in Manned Fighter Aircraft," *SETP Quarterly Review* 2, no. 3 (Fall 1957): 18–23.

42. Quoted in Mindell, *Between Human and Machine*, 77.

43. Harris, *Human Performance on the Flight Deck*, 221.

Chapter Four: THE DEGENERATION EFFECT

1. Alfred North Whitehead, *An Introduction to Mathematics* (New York: Henry Holt, 1911), 61.

2. Quoted in Frank Levy and Richard J. Murnane, *The New Division of Labor: How Computers Are Creating the Next Job Market* (Princeton: Princeton University Press, 2004), 4.

3. Raja Parasuraman et al., "Model for Types and Levels of Human Interaction with Automation," *IEEE Transactions on Systems, Man, and Cybernetics—Part A: Systems and Humans* 30, no. 3 (2000): 286–297. See also Nadine Sarter et al., "Automation Surprises," in Gavriel Salvendy, ed., *Handbook of Human Factors and Ergonomics*, 2nd ed. (New York: Wiley, 1997).

4. Dennis F. Galletta et al., "Does Spell-Checking Software Need a Warning Label?," *Communications of the ACM* 48, no. 7 (2005): 82–86.

5. National Transportation Safety Board, *Marine Accident Report: Grounding of the Panamanian Passenger Ship* Royal Majesty *on Rose and Crown Shoal near Nantucket, Massachusetts, June 10, 1995* (Washington, D.C.: NTSB, April 2, 1997).

6. Sherry Turkle, *Simulation and Its Discontents* (Cambridge, Mass.: MIT Press, 2009), 55–56.

7. Jennifer Langston, "GPS Routed Bus under Bridge, Company Says," *Seattle Post-Intelligencer*, April 17, 2008.

8. A. A. Povyakalo et al., "How to Discriminate between Computer-Aided and Computer-Hindered Decisions: A Case Study in Mammography," *Medical Decision Making* 33, no. 1 (January 2013): 98–107.

9. E. Alberdi et al., "Why Are People's Decisions Sometimes Worse with Computer Support?," in Bettina Buth et al., eds., *Proceedings of SAFE-COMP 2009, the 28th International Conference on Computer Safety, Reliability, and Security* (Hamburg, Germany: Springer, 2009), 18–31.

10. See Raja Parasuraman et al., "Performance Consequences of Automation-Induced 'Complacency,'" *International Journal of Aviation Psychology* 3, no. 1 (1993): 1–23.

11. Raja Parasuraman and Dietrich H. Manzey, "Complacency and Bias in Human Use of Automation: An Attentional Integration," *Human Factors* 52, no. 3 (June 2010): 381–410.

12. Norman J. Slamecka and Peter Graf, "The Generation Effect: Delineation of a Phenomenon," *Journal of Experimental Psychology: Human Learning and Memory* 4, no. 6 (1978): 592–604.

13. Jeffrey D. Karpicke and Janell R. Blunt, "Retrieval Practice Produces More Learning than Elaborative Studying with Concept Mapping," *Science* 331 (2011): 772–775.

14. Britte Haugan Cheng, "Generation in the Knowledge Integration Classroom" (PhD thesis, University of California, Berkeley, 2008).

15. Simon Farrell and Stephan Lewandowsky, "A Connectionist Model of Complacency and Adaptive Recovery under Automation," *Journal of Experimental Psychology: Learning, Memory, and Cognition* 26, no. 2 (2000): 395–410.

16. I first discussed van Nimwegen's work in my book *The Shallows: What the Internet Is Doing to Our Brains* (New York: W. W. Norton, 2010), 214–216.

17. Christof van Nimwegen, "The Paradox of the Guided User: Assistance Can Be Counter-effective" (SIKS Dissertation Series No. 2008-09, Utrecht University, March 31, 2008). See also Christof van Nimwegen and Herre van Oostendorp, "The Questionable Impact of an Assisting Interface on Performance in Transfer Situations," *International Journal of Industrial Ergonomics* 39, no. 3 (May 2009): 501–508; and Daniel Burgos and Christof van Nimwegen, "Games-Based Learning, Destination Feedback and Adaptation: A Case Study of an Educational Planning Simulation," in Thomas Connolly et al., eds., *Games-Based Learning Advancements for Multi-Sensory Human Computer Interfaces: Techniques and Effective Practices* (Hershey, Penn.: IGI Global, 2009), 119–130.

18. Carlin Dowling et al., "Audit Support System Design and the Declarative Knowledge of Long-Term Users," *Journal of Emerging Technologies in Accounting* 5, no. 1 (December 2008): 99–108.

19. See Richard G. Brody et al., "The Effect of a Computerized Decision Aid on the Development of Knowledge," *Journal of Business and Psychology* 18, no. 2 (2003): 157–174; and Holli McCall et al., "Use of Knowledge

Management Systems and the Impact on the Acquisition of Explicit Knowledge," *Journal of Information Systems* 22, no. 2 (2008): 77–101.

20. Amar Bhidé, "The Judgment Deficit," *Harvard Business Review* 88, no. 9 (September 2010): 44–53.

21. Gordon Baxter and John Cartlidge, "Flying by the Seat of Their Pants: What Can High Frequency Trading Learn from Aviation?," in G. Brat et al., eds., *ATACCS-2013: Proceedings of the 3rd International Conference on Application and Theory of Automation in Command and Control Systems* (New York: ACM, 2013), 64–73.

22. Vivek Haldar, "Sharp Tools, Dull Minds," *This Is the Blog of Vivek Haldar*, November 10, 2013, blog.vivekhaldar.com/post/66660163006/sharp-tools-dull-minds.

23. Tim Adams, "Google and the Future of Search: Amit Singhal and the Knowledge Graph," *Observer*, January 19, 2013.

24. Betsy Sparrow et al., "Google Effects on Memory: Cognitive Consequences of Having Information at Our Fingertips," *Science* 333, no. 6043 (August 5, 2011): 776–778. Another study suggests that simply knowing an experience has been photographed with a digital camera weakens a person's memory of the experience: Linda A. Henkel, "Point-and-Shoot Memories: The Influence of Taking Photos on Memory for a Museum Tour," *Psychological Science*, December 5, 2013, pss.sagepub.com/content/early/2013/12/04/0956797613504438.full.

25. Mihai Nadin, "Information and Semiotic Processes: The Semiotics of Computation," *Cybernetics and Human Knowing* 18, nos. 1–2 (2011): 153–175.

26. Gary Marcus, *Guitar Zero: The New Musician and the Science of Learning* (New York: Penguin, 2012), 52.

27. For a thorough description of how the brain learns to read, see Maryanne Wolf, *Proust and the Squid: The Story and Science of the Reading Brain* (New York: HarperCollins, 2007), particularly 108–133.

28. Hubert L. Dreyfus, "Intelligence without Representation—Merleau-Ponty's Critique of Mental Representation," *Phenomenology and the Cognitive Sciences* 1 (2002): 367–383.

29. Marcus, *Guitar Zero*, 103.

30. David Z. Hambrick and Elizabeth J. Meinz, "Limits on the Predictive Power of Domain-Specific Experience and Knowledge in Skilled Per-

formance," *Current Directions in Psychological Science* 20, no. 5 (2011): 275–279.

31. K. Anders Ericsson et al., "The Role of Deliberate Practice in the Acquisition of Expert Performance," *Psychological Review* 100, no. 3 (1993): 363–406.

32. Nigel Warburton, "Robert Talisse on Pragmatism," *Five Books*, September 18, 2013, fivebooks.com/interviews/robert-talisse-on-pragmatism.

33. Jeanne Nakamura and Mihaly Csikszentmihalyi, "The Concept of Flow," in C. R. Snyder and Shane J. Lopez, eds., *Handbook of Positive Psychology* (Oxford, U.K.: Oxford University Press, 2002), 90–91.

Interlude, with Dancing Mice

1. Robert M. Yerkes, *The Dancing Mouse: A Study in Animal Behavior* (New York: Macmillan, 1907), vii–viii, 2–3.

2. Ibid., vii.

3. Robert M. Yerkes and John D. Dodson, "The Relation of Strength of Stimulus to Rapidity of Habit-Formation," *Journal of Comparative Neurology and Psychology* 18 (1908): 459–482.

4. Ibid.

5. Mark S. Young and Neville A. Stanton, "Attention and Automation: New Perspectives on Mental Overload and Performance," *Theoretical Issues in Ergonomics Science* 3, no. 2 (2002): 178–194.

6. Mark W. Scerbo, "Adaptive Automation," in Raja Parasuraman and Matthew Rizzo, eds., *Neuroergonomics: The Brain at Work* (New York: Oxford University Press, 2007), 239–252.

Chapter Five: WHITE-COLLAR COMPUTER

1. "RAND Study Says Computerizing Medical Records Could Save $81 Billion Annually and Improve the Quality of Medical Care," RAND Corporation press release, September 14, 2005.

2. Richard Hillestad et al., "Can Electronic Medical Record Systems Transform Health Care? Potential Health Benefits, Savings, and Costs," *Health Affairs* 24, no. 5 (2005): 1103–1117.

3. Reed Abelson and Julie Creswell, "In Second Look, Few Savings from Digital Health Records," *New York Times*, January 10, 2013.

4. Jeanne Lambrew, "More than Half of Doctors Now Use Electronic Health Records Thanks to Administration Policies," *The White House Blog*, May 24, 2013, whitehouse.gov/blog/2013/05/24/more-half-doctors-use-electronic-health-records-thanks-administration-policies.

5. Arthur L. Kellermann and Spencer S. Jones, "What It Will Take to Achieve the As-Yet-Unfulfilled Promises of Health Information Technology," *Health Affairs* 32, no. 1 (2013): 63–68.

6. Ashly D. Black et al., "The Impact of eHealth on the Quality and Safety of Health Care: A Systematic Overview," *PLOS Medicine* 8, no. 1 (2011), plosmedicine.org/article/info%3Adoi%2F10.1371%2Fjournal.pmed.1000387.

7. Melinda Beeuwkes Buntin et al., "The Benefits of Health Information Technology: A Review of the Recent Literature Shows Predominantly Positive Results," *Health Affairs* 30, no. 3 (2011): 464–471.

8. Dean F. Sittig et al., "Lessons from 'Unexpected Increased Mortality after Implementation of a Commercially Sold Computerized Physician Order Entry System,'" *Pediatrics* 118, no. 2 (August 1, 2006): 797–801.

9. Jerome Groopman and Pamela Hartzband, "Obama's $80 Billion Exaggeration," *Wall Street Journal*, March 12, 2009. See also, by the same authors, "Off the Record—Avoiding the Pitfalls of Going Electronic," *New England Journal of Medicine* 358, no. 16 (2008): 1656–1658.

10. See Fred Schulte, "Growth of Electronic Medical Records Eases Path to Inflated Bills," Center for Public Integrity, September 19, 2012, publicintegrity.org/2012/09/19/10812/growth-electronic-medical-records-eases-path-inflated-bills; and Reed Abelson et al., "Medicare Bills Rise as Records Turn Electronic," *New York Times*, September 22, 2012.

11. Daniel R. Levinson, *CMS and Its Contractors Have Adopted Few Program Integrity Practices to Address Vulnerabilities in EHRs* (Washington, D.C.: Office of the Inspector General, Department of Health and Human Services, January 2014), oig.hhs.gov/oei/reports/oei-01-11-00571.pdf.

12. Danny McCormick et al., "Giving Office-Based Physicians Electronic Access to Patients' Prior Imaging and Lab Results Did Not Deter Ordering of Tests," *Health Affairs* 31, no. 3 (2012): 488–496. An earlier study tracked the treatment of diabetes patients over five years at two clinics, one that had installed an electronic medical record system and one that hadn't. It found that physicians at the clinic with the EMR system ordered more tests but did not achieve better glycemic control

in their patients. "The data suggest that despite the substantial cost and increasing technical sophistication of EMRs, EMR use failed to achieve desirable levels of clinical improvement," wrote the researchers. Patrick J. O'Connor et al., "Impact of an Electronic Medical Record on Diabetes Quality of Care," *Annals of Family Medicine* 3, no. 4 (July 2005): 300–306.

13. Timothy Hoff, "Deskilling and Adaptation among Primary Care Physicians Using Two Work Innovations," *Health Care Management Review* 36, no. 4 (2011): 338–348.

14. Schulte, "Growth of Electronic Medical Records."

15. Hoff, "Deskilling and Adaptation."

16. Danielle Ofri, "The Doctor vs. the Computer," *New York Times*, December 30, 2010.

17. Thomas H. Payne et al., "Transition from Paper to Electronic Inpatient Physician Notes," *Journal of the American Medical Information Association* 17 (2010): 108–111.

18. Ofri, "Doctor vs. the Computer."

19. Beth Lown and Dayron Rodriguez, "Lost in Translation? How Electronic Health Records Structure Communication, Relationships, and Meaning," *Academic Medicine* 87, no. 4 (2012): 392–394.

20. Emran Rouf et al., "Computers in the Exam Room: Differences in Physician-Patient Interaction May Be Due to Physician Experience," *Journal of General Internal Medicine* 22, no. 1 (2007): 43–48.

21. Avik Shachak et al., "Primary Care Physicians' Use of an Electronic Medical Record System: A Cognitive Task Analysis," *Journal of General Internal Medicine* 24, no. 3 (2009): 341–348.

22. Lown and Rodriguez, "Lost in Translation?"

23. See Saul N. Weingart et al., "Physicians' Decisions to Override Computerized Drug Alerts in Primary Care," *Archives of Internal Medicine* 163 (November 24, 2003): 2625–2631; Alissa L. Russ et al., "Prescribers' Interactions with Medication Alerts at the Point of Prescribing: A Multi-method, *In Situ* Investigation of the Human–Computer Interaction," *International Journal of Medical Informatics* 81 (2012): 232–243; M. Susan Ridgely and Michael D. Greenberg, "Too Many Alerts, Too Much Liability: Sorting through the Malpractice Implications of Drug-Drug Interaction Clinical Decision Support," *Saint Louis University Journal of Health Law and Policy* 5 (2012): 257–295;

and David W. Bates, "Clinical Decision Support and the Law: The Big Picture," *Saint Louis University Journal of Health Law and Policy* 5 (2012): 319–324.

24. Atul Gawande, *The Checklist Manifesto: How to Get Things Right* (New York: Henry Holt, 2010), 161–162.

25. Lown and Rodriguez, "Lost in Translation?"

26. Jerome Groopman, *How Doctors Think* (New York: Houghton Mifflin, 2007), 34–35.

27. Adam Smith, *The Wealth of Nations* (New York: Modern Library, 2000), 840.

28. Ibid., 4.

29. Frederick Winslow Taylor, *The Principles of Scientific Management* (New York: Harper & Brothers, 1913), 11.

30. Ibid., 36.

31. Hannah Arendt, *The Human Condition* (Chicago: University of Chicago Press, 1998), 147.

32. Harry Braverman, *Labor and Monopoly Capital: The Degradation of Work in the Twentieth Century* (New York: Monthly Review Press, 1998), 307.

33. For a succinct review of the Braverman debate, see Peter Meiksins, "Labor and Monopoly Capital for the 1990s: A Review and Critique of the Labor Process Debate," *Monthly Review*, November 1994.

34. James R. Bright, *Automation and Management* (Cambridge, Mass.: Harvard University, 1958), 176–195.

35. Ibid., 188.

36. James R. Bright, "The Relationship of Increasing Automation and Skill Requirements," in National Commission on Technology, Automation, and Economic Progress, *Technology and the American Economy, Appendix II: The Employment Impact of Technological Change* (Washington, D.C.: U.S. Government Printing Office, 1966), 201–221.

37. George Dyson, comment on Edge.org, July 11, 2008, edge.org/discourse/carr_google.html#dysong.

38. For a lucid explanation of machine learning, see the sixth chapter of John MacCormick's *Nine Algorithms That Changed the Future: The Ingenious Ideas That Drive Today's Computers* (Princeton: Princeton University Press, 2012).

39. Max Raskin and Ilan Kolet, "Wall Street Jobs Plunge as Profits Soar," Bloomberg News, April 23, 2013, bloomberg.com/news/2013-04-24/wall-street-jobs-plunge-as-profits-soar-chart-of-the-day.html.

40. Ashwin Parameswaran, "Explaining the Neglect of Doug Engelbart's Vision: The Economic Irrelevance of Human Intelligence Augmentation," Macroresilience, July 8, 2013, macroresilience.com/2013/07/08/explaining-the-neglect-of-doug-engelbarts-vision/.

41. See Daniel Martin Katz, "Quantitative Legal Prediction—or—How I Learned to Stop Worrying and Start Preparing for the Data-Driven Future of the Legal Services Industry," Emory Law Journal 62, no. 4 (2013): 909–966.

42. Joseph Walker, "Meet the New Boss: Big Data," Wall Street Journal, September 20, 2012.

43. Franco "Bifo" Berardi, The Soul at Work: From Alienation to Automation (Los Angeles: Semiotext(e), 2009), 96.

44. A. M. Turing, "Systems of Logic Based on Ordinals," Proceedings of the London Mathematical Society 45, no. 2239 (1939): 161–228.

45. Ibid.

46. Hector J. Levesque, "On Our Best Behaviour," lecture delivered at the International Joint Conference on Artificial Intelligence, Beijing, China, August 8, 2013.

47. See Nassim Nicholas Taleb, Antifragile: Things That Gain from Disorder (New York: Random House, 2012), 416–419.

48. Donald T. Campbell, "Assessing the Impact of Planned Social Change," Occasional Paper Series, no. 8 (December 1976), Public Affairs Center, Dartmouth College, Hanover, N.H.

49. Viktor Mayer-Schönberger and Kenneth Cukier, Big Data: A Revolution That Will Transform How We Live, Work, and Think (New York: Houghton Mifflin Harcourt, 2013), 166.

50. Kate Crawford, "The Hidden Biases in Big Data," HBR Blog Network, April 1, 2013, hbr.org/cs/2013/04/the_hidden_biases_in_big_data.html.

51. In a 1968 article, Weed wrote, "If useful historical data can be acquired and stored cheaply, completely and accurately by new computers and interviewing technics without the use of expensive physician time, they should be seriously considered." Lawrence L. Weed, "Medical Records That Guide and Teach," New England Journal of Medicine 278 (1968): 593–600, 652–657.

52. Lee Jacobs, "Interview with Lawrence Weed, MD—The Father of the Problem-Oriented Medical Record Looks Ahead," *Permanente Journal* 13, no. 3 (2009): 84–89.

53. Gary Klein, "Evidence-Based Medicine," *Edge*, January 14, 2014, edge .org/responses/what-scientific-idea-is-ready-for-retirement.

54. Michael Oakeshott, "Rationalism in Politics," *Cambridge Journal* 1 (1947): 81–98, 145–157. The essay was collected in Oakeshott's 1962 book *Rationalism in Politics and Other Essays* (New York: Basic Books).

Chapter Six: WORLD AND SCREEN

1. William Edward Parry, *Journal of a Second Voyage for the Discovery of a North-West Passage from the Atlantic to the Pacific* (London: John Murray, 1824), 277.

2. Claudio Aporta and Eric Higgs, "Satellite Culture: Global Positioning Systems, Inuit Wayfinding, and the Need for a New Account of Technology," *Current Anthropology* 46, no. 5 (2005): 729–753.

3. Interview of Claudio Aporta by author, January 25, 2012.

4. Gilly Leshed et al., "In-Car GPS Navigation: Engagement with and Disengagement from the Environment," in *Proceedings of the SIGCHI Conference on Human Factors in Computing Systems* (New York: ACM, 2008), 1675–1684.

5. David Brooks, "The Outsourced Brain," *New York Times*, October 26, 2007.

6. Julia Frankenstein et al., "Is the Map in Our Head Oriented North?," *Psychological Science* 23, no. 2 (2012): 120–125.

7. Julia Frankenstein, "Is GPS All in Our Heads?," *New York Times*, February 2, 2012.

8. Gary E. Burnett and Kate Lee, "The Effect of Vehicle Navigation Systems on the Formation of Cognitive Maps," in Geoffrey Underwood, ed., *Traffic and Transport Psychology: Theory and Application* (Amsterdam: Elsevier, 2005), 407–418.

9. Elliot P. Fenech et al., "The Effects of Acoustic Turn-by-Turn Navigation on Wayfinding," *Proceedings of the Human Factors and Ergonomics Society Annual Meeting* 54, no. 23 (2010): 1926–1930.

10. Toru Ishikawa et al., "Wayfinding with a GPS-Based Mobile Navigation System: A Comparison with Maps and Direct Experience," *Journal of*

Environmental Psychology 28, no. 1 (2008): 74–82; and Stefan Münzer et al., "Computer-Assisted Navigation and the Acquisition of Route and Survey Knowledge," *Journal of Environmental Psychology* 26, no. 4 (2006): 300–308.

11. Sara Hendren, "The White Cane as Technology," *Atlantic*, November 6, 2013, theatlantic.com/technology/archive/2013/11/the-white-cane-as-technology/281167/.

12. Tim Ingold, *Being Alive: Essays on Movement, Knowledge and Description* (London: Routledge, 2011), 149–152. The emphasis is Ingold's.

13. Quoted in James Fallows, "The Places You'll Go," *Atlantic*, January/February 2013.

14. Ari N. Schulman, "GPS and the End of the Road," *New Atlantis*, Spring 2011.

15. John O'Keefe and Jonathan Dostrovsky, "The Hippocampus as a Spatial Map: Preliminary Evidence from Unit Activity in the Freely-Moving Rat," *Brain Research* 34 (1971): 171–175.

16. John O'Keefe, "A Review of the Hippocampal Place Cells," *Progress in Neurobiology* 13, no. 4 (2009): 419–439.

17. Edvard I. Moser et al., "Place Cells, Grid Cells, and the Brain's Spatial Representation System," *Annual Review of Neuroscience* 31 (2008): 69–89.

18. See Christian F. Doeller et al., "Evidence for Grid Cells in a Human Memory Network," *Nature* 463 (2010): 657–661; Nathaniel J. Killian et al., "A Map of Visual Space in the Primate Entorhinal Cortex," *Nature* 491 (2012): 761–764; and Joshua Jacobs et al., "Direct Recordings of Grid-Like Neuronal Activity in Human Spatial Navigation," *Nature Neuroscience,* August 4, 2013, nature.com/neuro/journal/vaop/ncurrent/full/nn.3466.html.

19. James Gorman, "A Sense of Where You Are," *New York Times*, April 30, 2013.

20. György Buzsáki and Edvard I. Moser, "Memory, Navigation and Theta Rhythm in the Hippocampal-Entorhinal System," *Nature Neuroscience* 16, no. 2 (2013): 130–138. See also Neil Burgess et al., "Memory for Events and Their Spatial Context: Models and Experiments," in Alan Baddeley et al., eds., *Episodic Memory: New Directions in Research* (New York: Oxford University Press, 2002), 249–268. It seems revealing that one of the most powerful mnemonic devices, dating back to classical

times, involves setting mental pictures of items or facts in locations in an imaginary place, such as a building or a town. Memories become easier to recall when they're associated with physical locations, even if only in the imagination.

21. See, for example, Jan M. Wiener et al., "Maladaptive Bias for Extrahippocampal Navigation Strategies in Aging Humans," *Journal of Neuroscience* 33, no. 14 (2013): 6012–6017.

22. See, for example, A. T. Du et al., "Magnetic Resonance Imaging of the Entorhinal Cortex and Hippocampus in Mild Cognitive Impairment and Alzheimer's Disease," *Journal of Neurology, Neurosurgery and Psychiatry* 71 (2001): 441–447.

23. Kyoko Konishi and Véronique D. Bohbot, "Spatial Navigational Strategies Correlate with Gray Matter in the Hippocampus of Healthy Older Adults Tested in a Virtual Maze," *Frontiers in Aging Neuroscience* 5 (2013): 1–8.

24. Email from Véronique Bohbot to author, June 4, 2010.

25. Quoted in Alex Hutchinson, "Global Impositioning Systems," *Walrus*, November 2009.

26. Kyle VanHemert, "4 Reasons Why Apple's iBeacon Is About to Disrupt Interaction Design," *Wired*, December 11, 2013, www.wired.com/design/2013/12/4-use-cases-for-ibeacon-the-most-exciting-tech-you-havent-heard-of/.

27. Quoted in Fallows, "Places You'll Go."

28. Damon Lavrinc, "Mercedes Is Testing Google Glass Integration, and It Actually Works," *Wired*, August 15, 2013, wired.com/autopia/2013/08/google-glass-mercedes-benz/.

29. William J. Mitchell, "Foreword," in Yehuda E. Kalay, *Architecture's New Media: Principles, Theories, and Methods of Computer-Aided Design* (Cambridge, Mass.: MIT Press, 2004), xi.

30. Anonymous, "Interviews: Renzo Piano," *Architectural Record*, October 2001, archrecord.construction.com/people/interviews/archives/0110piano.asp.

31. Quoted in Gavin Mortimer, *The Longest Night* (New York: Penguin, 2005), 319.

32. Dino Marcantonio, "Architectural Quackery at Its Finest: Parametricism," *Marcantonio Architects Blog*, May 8, 2010, blog.marcantonioarchitects.com/architectural-quackery-at-its-finest-parametricism/.

33. Paul Goldberger, "Digital Dreams," *New Yorker*, March 12, 2001.

34. Patrik Schumacher, "Parametricism as Style—Parametricist Manifesto," Patrik Schumacher's blog, 2008, patrikschumacher.com/Texts/Parametricism%20as%20Style.htm.

35. Anonymous, "Interviews: Renzo Piano."

36. Witold Rybczynski, "Think before You Build," *Slate*, March 30, 2011, slate.com/articles/arts/architecture/2011/03/think_before_you_build.html.

37. Quoted in Bryan Lawson, *Design in Mind* (Oxford, U.K.: Architectural Press, 1994), 66.

38. Michael Graves, "Architecture and the Lost Art of Drawing," *New York Times*, September 2, 2012.

39. D. A. Schön, "Designing as Reflective Conversation with the Materials of a Design Situation," *Knowledge-Based Systems* 5, no. 1 (1992): 3–14. See also Schön's book *The Reflective Practitioner: How Professionals Think in Action* (New York: Basic Books, 1983), particularly 157–159.

40. Graves, "Architecture and the Lost Art of Drawing." See also Masaki Suwa et al., "Macroscopic Analysis of Design Processes Based on a Scheme for Coding Designers' Cognitive Actions," *Design Studies* 19 (1998): 455–483.

41. Nigel Cross, *Designerly Ways of Knowing* (Basel: Birkhäuser, 2007), 58.

42. Schön, "Designing as Reflective Conversation."

43. Ibid.

44. Joachim Walther et al., "Avoiding the Potential Negative Influence of CAD Tools on the Formation of Students' Creativity," in *Proceedings of the 2007 AaeE Conference*, Melbourne, Australia, December 2007, ww2.cs.mu.oz.au/aaee2007/papers/paper_40.pdf.

45. Graves, "Architecture and the Lost Art of Drawing."

46. Juhani Pallasmaa, *The Thinking Hand: Existential and Embodied Wisdom in Architecture* (Chichester, U.K.: Wiley, 2009), 96–97.

47. Interview of E. J. Meade by author, July 23, 2013.

48. Jacob Brillhart, "Drawing towards a More Creative Architecture: Mediating between the Digital and the Analog," paper presented at the annual meeting of the Association of Collegiate Schools of Architecture, Montreal, Canada, March 5, 2011.

49. Matthew B. Crawford, *Shop Class as Soulcraft: An Inquiry into the Value of Work* (New York: Penguin, 2009), 164.

50. Ibid., 161.

51. John Dewey, *Essays in Experimental Logic* (Chicago: University of Chicago Press, 1916), 13–14.

52. Matthew D. Lieberman, "The Mind-Body Illusion," *Psychology Today*, May 17, 2012, psychologytoday.com/blog/social-brain-social-mind/201205/the-mind-body-illusion. See also Matthew D. Lieberman, "What Makes Big Ideas Sticky?," in Max Brockman, ed., *What's Next? Dispatches on the Future of Science* (New York: Vintage, 2009), 90–103.

53. "Andy Clark: Embodied Cognition" (video), University of Edinburgh: Research in a Nutshell, undated, nutshell-videos.ed.ac.uk/andy-clark-embodied-cognition.

54. Tim Gollisch and Markus Meister, "Eye Smarter than Scientists Believed: Neural Computations in Circuits of the Retina," *Neuron* 65 (January 28, 2010): 150–164.

55. See Vittorio Gallese and George Lakoff, "The Brain's Concepts: The Role of the Sensory-Motor System in Conceptual Knowledge," *Cognitive Neuropsychology* 22, no. 3/4 (2005): 455–479; and Lawrence W. Barsalou, "Grounded Cognition," *Annual Review of Psychology* 59 (2008): 617–645.

56. "Andy Clark: Embodied Cognition."

57. Shaun Gallagher, *How the Body Shapes the Mind* (Oxford, U.K.: Oxford University Press, 2005), 247.

58. Andy Clark, *Natural-Born Cyborgs: Minds, Technologies, and the Future of Human Intelligence* (New York: Oxford University Press, 2003), 4.

59. Quoted in Fallows, "Places You'll Go."

Chapter Seven: AUTOMATION FOR THE PEOPLE

1. Kevin Kelly, "Better than Human: Why Robots Will—and Must—Take Our Jobs," *Wired*, January 2013.

2. Jay Yarow, "Human Driver Crashes Google's Self Driving Car," *Business Insider*, August 5, 2011, businessinsider.com/googles-self-driving-cars-get-in-their-first-accident-2011-8.

3. Andy Kessler, "Professors Are About to Get an Online Education," *Wall Street Journal*, June 3, 2013.

4. Vinod Khosla, "Do We Need Doctors or Algorithms?," TechCrunch, January 10, 2012, techcrunch.com/2012/01/10/doctors-or-algorithms.

5. Gerald Traufetter, "The Computer vs. the Captain: Will Increasing Automation Make Jets Less Safe?," *Spiegel Online*, July 31, 2009, spiegel .de/international/world/the-computer-vs-the-captain-will-increasing-automation-make-jets-less-safe-a-639298.html.

6. See Adam Fisher, "Inside Google's Quest to Popularize Self-Driving Cars," *Popular Science*, October 2013.

7. Tosha B. Weeterneck et al., "Factors Contributing to an Increase in Duplicate Medication Order Errors after CPOE Implementation," *Journal of the American Medical Informatics Association* 18 (2011): 774–782.

8. Sergey V. Buldyrev et al., "Catastrophic Cascade of Failures in Interdependent Networks," *Nature* 464 (April 15, 2010): 1025–1028. See also Alessandro Vespignani, "The Fragility of Interdependency," *Nature* 464 (April 15, 2010): 984–985.

9. Nancy G. Leveson, *Engineering a Safer World: Systems Thinking Applied to Safety* (Cambridge, Mass.: MIT Press, 2011), 8–9.

10. Lisanne Bainbridge, "Ironies of Automation," *Automatica* 19, no. 6 (1983): 775–779.

11. For a review of research on vigilance, including the World War II studies, see D. R. Davies and R. Parasuraman, *The Psychology of Vigilance* (London: Academic Press, 1982).

12. Bainbridge, "Ironies of Automation."

13. See Magdalen Galley, "Ergonomics—Where Have We Been and Where Are We Going," undated speech, taylor.it/meg/papers/50%20Years%20 of%20Ergonomics.pdf; and Nicolas Marmaras et al., "Ergonomic Design in Ancient Greece," *Applied Ergonomics* 30, no. 4 (1999): 361–368.

14. David Meister, *The History of Human Factors and Ergonomics* (Mahwah, N.J.: Lawrence Erlbaum Associates, 1999), 209, 359.

15. Leo Marx, "Does Improved Technology Mean Progress?," *Technology Review*, January 1987.

16. Donald A. Norman, *Things That Make Us Smart: Defending Human Attributes in the Age of the Machine* (New York: Perseus, 1993), xi.

17. Norbert Wiener, *I Am a Mathematician* (Cambridge, Mass.: MIT Press, 1956), 305.

18. Nadine Sarter et al., "Automation Surprises," in Gavriel Salvendy, ed., *Handbook of Human Factors and Ergonomics*, 2nd ed. (New York: Wiley, 1997).

19. Ibid.

20. John D. Lee, "Human Factors and Ergonomics in Automation Design," in Gavriel Salvendy, ed., *Handbook of Human Factors and Ergonomics*, 3rd ed. (Hoboken, N.J.: Wiley, 2006), 1571.

21. For more on human-centered automation, see Charles E. Billings, *Aviation Automation: The Search for a Human-Centered Approach* (Mahwah, N.J.: Lawrence Erlbaum Associates, 1997); and Raja Parasuraman et al., "A Model for Types and Levels of Human Interaction with Automation," *IEEE Transactions on Systems, Man, and Cybernetics* 30, no. 3 (2000): 286–297.

22. David B. Kaber et al., "On the Design of Adaptive Automation for Complex Systems," *International Journal of Cognitive Ergonomics* 5, no. 1 (2001): 37–57.

23. Mark W. Scerbo, "Adaptive Automation," in Raja Parasuraman and Matthew Rizzo, eds., *Neuroergonomics: The Brain at Work* (New York: Oxford University Press, 2007), 239–252. For more on the DARPA project, see Mark St. John et al., "Overview of the DARPA Augmented Cognition Technical Integration Experiment," *International Journal of Human-Computer Interaction* 17, no. 2 (2004): 131–149.

24. Lee, "Human Factors and Ergonomics."

25. Interview of Raja Parasuraman by author, December 18, 2011.

26. Lee, "Human Factors and Ergonomics."

27. Interview of Ben Tranel by author, June 13, 2013.

28. Mark D. Gross and Ellen Yi-Luen Do, "Ambiguous Intentions: A Paperlike Interface for Creative Design," in *Proceedings of the ACM Symposium on User Interface Software and Technology* (New York: ACM, 1996), 183–192.

29. Julie Dorsey et al., "The Mental Canvas: A Tool for Conceptual Architectural Design and Analysis," in *Proceedings of the Pacific Conference on Computer Graphics and Applications* (2007), 201–210.

30. William Langewiesche, *Fly by Wire: The Geese, the Glide, the "Miracle" on the Hudson* (New York: Farrar, Straus & Giroux, 2009), 102.

31. Lee, "Human Factors and Ergonomics."

32. CBS News, "Faulty Data Misled Pilots in '09 Air France Crash," July 5, 2012, cbsnews.com/8301-505263_162-57466644/faulty-data-misled-pilots-in-09-air-france-crash/.

33. Langewiesche, *Fly by Wire*, 109.

34. Federal Aviation Administration, "NextGen Air Traffic Control/ Technical Operations Human Factors (Controller Efficiency & Air Ground Integration) Research and Development Plan," version one, April 2011.

35. Nathaniel Popper, "Bank Gains by Putting Brakes on Traders," *New York Times*, June 26, 2013.

36. Thomas P. Hughes, "Technological Momentum," in Merritt Roe Smith and Leo Marx, eds., *Does Technology Drive History? The Dilemma of Technological Determinism* (Cambridge, Mass.: MIT Press, 1994), 101–113.

37. Gordon Baxter and John Cartlidge, "Flying by the Seat of Their Pants: What Can High Frequency Trading Learn from Aviation?," in G. Brat et al., eds., *ATACCS-2013: Proceedings of the 3rd International Conference on Application and Theory of Automation in Command and Control Systems* (New York: ACM, 2013), 64–73.

38. David F. Noble, *Forces of Production: A Social History of Industrial Automation* (New York: Alfred A. Knopf, 1984), 144–145.

39. Ibid., 94.

40. Quoted in Noble, *Forces of Production*, 94.

41. Ibid., 326.

42. Dyson made this comment in the 1981 documentary *The Day after Trinity*. Quoted in Bill Joy, "Why the Future Doesn't Need Us," *Wired*, April 2000.

43. Matt Richtel, "A Silicon Valley School That Doesn't Compute," *New York Times*, October 23, 2011.

Interlude, with Grave Robber

1. Peter Merholz, "'Frictionless' as an Alternative to 'Simplicity' in Design," *Adaptive Path* (blog), July 21, 2010, adaptivepath.com/ideas/ friction-as-an-alternative-to-simplicity-in-design.

2. David J. Hill, "Exclusive Interview with Ray Kurzweil on Future AI Project at Google," *SingularityHUB*, January 10, 2013, singularityhub .com/2013/01/10/exclusive-interview-with-ray-kurzweil-on-future-ai-project-at-google/.

Chapter Eight: YOUR INNER DRONE

1. Asimov's Rules of Robotics—"the three rules that are built most deeply into a robot's positronic brain"—first appeared in his 1942 short story "Runaround," which can be found in the collection *I, Robot* (New York: Bantam, 2004), 37.

2. Gary Marcus, "Moral Machines," *News Desk* (blog), *New Yorker*, November 27, 2012, newyorker.com/online/blogs/newsdesk/2012/11/google-dri verless-car-morality.html.

3. Charles T. Rubin, "Machine Morality and Human Responsibility," *New Atlantis*, Summer 2011.

4. Christof Heyns, "Report of the Special Rapporteur on Extrajudicial, Summary or Arbitrary Executions," presentation to the Human Rights Council of the United Nations General Assembly, April 9, 2013, www.ohchr.org/Documents/HRBodies/HRCouncil/RegularSession/Session23/A-HRC-23-47_en.pdf.

5. Patrick Lin et al., "Autonomous Military Robotics: Risk, Ethics, and Design," version 1.0.9, prepared for U.S. Department of Navy, Office of Naval Research, December 20, 2008.

6. Ibid.

7. Thomas K. Adams, "Future Warfare and the Decline of Human Decisionmaking," *Parameters*, Winter 2001–2002.

8. Heyns, "Report of the Special Rapporteur."

9. Ibid.

10. Joseph Weizenbaum, *Computer Power and Human Reason: From Judgment to Calculation* (New York: W. H. Freeman, 1976), 20.

11. Mark Weiser, "The Computer for the 21st Century," *Scientific American*, September 1991.

12. Mark Weiser and John Seely Brown, "The Coming Age of Calm Technology," in P. J. Denning and R. M. Metcalfe, eds., *Beyond Calculation: The Next Fifty Years of Computing* (New York: Springer, 1997), 75–86.

13. M. Weiser et al., "The Origins of Ubiquitous Computing Research at PARC in the Late 1980s," *IBM Systems Journal* 38, no. 4 (1999): 693–696.

14. See Nicholas Carr, *The Big Switch: Rewiring the World, from Edison to Google* (New York: W. W. Norton, 2008).

15. Thomas P. Hughes, *Networks of Power: Electrification in Western Society, 1880–1930* (Baltimore: Johns Hopkins University Press, 1983), 140.

16. W. Brian Arthur, "The Second Economy," *McKinsey Quarterly*, October 2011.

17. Ibid.

18. Bill Gates, *Business @ the Speed of Thought: Using a Digital Nervous System* (New York: Warner Books, 1999), 37.

19. Arthur C. Clarke, *Profiles of the Future: An Inquiry into the Limits of the Possible* (New York: Harper & Row, 1960), 227.

20. Sergey Brin, "Why Google Glass?," speech at TED2013, Long Beach, Calif., February 27, 2013, youtube.com/watch?v=rie-hPVJ7Sw.

21. Ibid.

22. See Christopher D. Wickens and Amy L. Alexander, "Attentional Tunneling and Task Management in Synthetic Vision Displays," *International Journal of Aviation Psychology* 19, no. 2 (2009): 182–199.

23. Richard F. Haines, "A Breakdown in Simultaneous Information Processing," in Gerard Obrecht and Lawrence W. Stark, eds., *Presbyopia Research: From Molecular Biology to Visual Adaptation* (New York: Plenum Press, 1991), 171–176.

24. Daniel J. Simons and Christopher F. Chambris, "Is Google Glass Dangerous?," *New York Times*, May 26, 2013.

25. "Amanda Rosenberg: Google Co-Founder Sergey Brin's New Girlfriend?," *Guardian*, August 30, 2013, theguardian.com/technology/shortcuts/2013/aug/30/amanda-rosenberg-google-sergey-brin-girlfriend.

26. Weiser, "Computer for the 21st Century."

27. Interview with Charlie Rose, *Charlie Rose*, April 24, 2012, charlierose.com/watch/60065884.

28. David Kirkpatrick, *The Facebook Effect* (New York: Simon & Schuster, 2010), 10.

29. Josh Constine, "Google Unites Gmail and G+ Chat into 'Hangouts' Cross-Platform Text and Group Video Messaging App," *TechCrunch*, May 15, 2013, techcrunch.com/2013/05/15/google-hangouts-messaging-app/.

30. Larry Greenemeier, "Chipmaker Races to Save Stephen Hawking's Speech as His Condition Deteriorates," *Scientific American*, January 18, 2013, www.scientificamerican.com/article.cfm?id=intel-helps-hawking-communicate.

31. Nick Bilton, "Disruptions: Next Step for Technology Is Becoming the Background," *New York Times*, July 1, 2012, bits.blogs.nytimes.com/2012/07/01/google's-project-glass-lets-technology-slip-into-the-background/.

32. Bruno Latour, "Morality and Technology: The End of the Means," *Theory, Culture and Society* 19 (2002): 247–260. The emphasis is Latour's.

33. Bernhard Seefeld, "Meet the New Google Maps: A Map for Every Person and Place," *Google Lat Long* (blog), May 15, 2013, google-latlong.blogspot.com/2013/05/meet-new-google-maps-map-for-every.html.

34. Evgeny Morozov, "My Map or Yours?," *Slate*, May 28, 2013, slate.com/articles/technology/future_tense/2013/05/google_maps_personalization_will_hurt_public_space_and_engagement.html.

35. Kirkpatrick, *Facebook Effect*, 199.

36. Sebastian Thrun, "Google's Driverless Car," speech at TED2011, March 2011, ted.com/talks/sebastian_thrun_google_s_driverless_car.html.

37. National Safety Council, "Annual Estimate of Cell Phone Crashes 2012," white paper, 2014.

38. See Sigfried Giedion, *Mechanization Takes Command* (New York: Oxford University Press, 1948), 628–712.

39. Langdon Winner, *Autonomous Technology: Technics-out-of-Control as a Theme in Political Thought* (Cambridge, Mass.: MIT Press, 1977), 285.

Chapter Nine: THE LOVE THAT LAYS THE SWALE IN ROWS

1. Quoted in Richard Poirier, *Robert Frost: The Work of Knowing* (Stanford, Calif.: Stanford University Press, 1990), 30. Details about Frost's life are drawn from Poirier's book; William H. Pritchard, *Frost: A Literary Life Reconsidered* (New York: Oxford University Press, 1984); and Jay Parini, *Robert Frost: A Life* (New York: Henry Holt, 1999).

2. Quoted in Poirier, *Robert Frost*, 30.

3. Robert Frost, "Mowing," in *A Boy's Will* (New York: Henry Holt, 1915), 36.

4. Robert Frost, "Two Tramps in Mud Time," in *A Further Range* (New York: Henry Holt, 1936), 16–18.

5. Poirier, *Robert Frost*, 278.

6. Robert Frost, "Some Science Fiction," in *In the Clearing* (New York: Holt, Rinehart & Winston, 1962), 89–90.

7. Poirier, *Robert Frost*, 301.

8. Robert Frost, "Kitty Hawk," in *In the Clearing*, 41–58.

9. Maurice Merleau-Ponty, *Phenomenology of Perception* (London: Routledge, 2012), 147. My reading of Merleau-Ponty draws on Hubert L. Dreyfus's commentary "The Current Relevance of Merleau-Ponty's Phenomenology of Embodiment," *Electronic Journal of Analytic Philosophy* 4 (Spring 1996), ejap.louisiana.edu/ejap/1996.spring/dreyfus.1996.spring.html.

10. Benedict de Spinoza, *Ethics* (London: Penguin, 1996), 44.

11. John Edward Huth, "Losing Our Way in the World," *New York Times*, July 21, 2013. See also Huth's enlightening book *The Lost Art of Finding Our Way* (Cambridge, Mass.: Harvard University Press, 2013).

12. Merleau-Ponty, *Phenomenology of Perception*, 148.

13. Ibid., 261.

14. See Nicholas Carr, *The Shallows: What the Internet Is Doing to Our Brains* (New York: W. W. Norton, 2010).

15. Pascal Ravassard et al., "Multisensory Control of Hippocampal Spatiotemporal Selectivity," *Science* 340, no. 6138 (2013): 1342–1346.

16. Anonymous, "Living in *The Matrix* Requires Less Brain Power," *Science Now*, May 2, 2013, news.sciencemag.org/physics/2013/05/living-matrix-requires-less-brain-power.

17. Alfred Korzybski, *Science and Sanity: An Introduction to Non-Aristotelian Systems and General Semantics*, 5th ed. (New York: Institute of General Semantics, 1994), 58.

18. John Dewey, *Art as Experience* (New York: Perigee Books, 1980), 59.

19. Medco, "America's State of Mind," 2011, apps.who.int/medicinedocs/documents/s19032en/s19032en.pdf.

20. Erin M. Sullivan et al., "Suicide among Adults Aged 35–64 Years—United States, 1999–2010," *Morbidity and Mortality Weekly Report*, May 3, 2013.

21. Alan Schwarz and Sarah Cohen, "A.D.H.D. Seen in 11% of U.S. Children as Diagnoses Rise," *New York Times*, April 1, 2013.

22. Robert Frost, "The Tuft of Flowers," in *A Boy's Will*, 47–49.

23. See Anonymous, "Fields of Automation," *Economist*, December 10, 2009; and Ian Berry, "Teaching Drones to Farm," *Wall Street Journal*, September 20, 2011.

24. Charles A. Lindbergh, *The Spirit of St. Louis* (New York: Scribner, 2003), 486. The emphasis is Lindbergh's.

25. J. C. R. Licklider, "Man-Computer Symbiosis," *IRE Transactions on Human Factors in Electronics* 1 (March 1960): 4–11.

26. Langdon Winner, *Autonomous Technology: Technics-out-of-Control as a Theme in Political Thought* (Cambridge, Mass.: MIT Press, 1977), 20–21.

27. Aristotle, *The Politics*, in Mitchell Cohen and Nicole Fermon, eds., *Princeton Readings in Political Thought* (Princeton: Princeton University Press, 1996), 110–111.

28. Evgeny Morozov, *To Save Everything, Click Here: The Folly of Technological Solutionism* (New York: PublicAffairs, 2013), 323.

29. Kevin Kelly, "Better than Human: Why Robots Will—and Must—Take Our Jobs," *Wired*, January 2013.

30. Kevin Drum, "Welcome, Robot Overloads. Please Don't Fire Us?," *Mother Jones*, May/June 2013.

31. Karl Marx and Frederick Engels, *The Communist Manifesto* (New York: Verso, 1998), 43.

32. Anonymous, "Slaves to the Smartphone," *Economist*, March 10, 2012.

33. Kevin Kelly, "What Technology Wants," *Cool Tools*, October 18, 2010, kk.org/cooltools/archives/4749.

34. George Packer, "No Death, No Taxes," *New Yorker*, November 28, 2011.

35. Hannah Arendt, *The Human Condition* (Chicago: University of Chicago Press, 1998), 4–5.

36. Mihaly Csikszentmihalyi, *Flow: The Psychology of Optimal Experience* (New York: Harper, 1991), 80.

37. Ralph Waldo Emerson, "The American Scholar," in *Essays and Lectures* (New York: Library of America, 1983), 57.

ACKNOWLEDGMENTS

The epigraph to this book is the concluding stanza of William Carlos Williams's poem "To Elsie," which appeared in the 1923 volume *Spring and All*.

I am deeply grateful to those who, as interviewees, reviewers, or correspondents, provided me with insight and assistance: Claudio Aporta, Henry Beer, Véronique Bohbot, George Dyson, Gerhard Fischer, Mark Gross, Katherine Hayles, Charles Jacobs, Joan Lowy, E. J. Meade, Raja Parasuraman, Lawrence Port, Jeff Robbins, Jeffrey Rowe, Ari Schulman, Evan Selinger, Betsy Sparrow, Tim Swan, Ben Tranel, and Christof van Nimwegen.

The Glass Cage is the third of my books to have been guided by the editorial hand of Brendan Curry at W. W. Norton. I thank Brendan and his colleagues for their work on my behalf. I am indebted as well to my agent, John Brockman, and his associates at Brockman Inc. for their wise counsel and support.

Some passages in this book appeared earlier, in different forms, in the *Atlantic*, the *Washington Post*, *MIT Technology Review*, and my blog, Rough Type.

INDEX